BEYOND THE CAUSAL HORIZON

Hawking Radiation, Other Energy Extractions from Black Holes, and Penrose Diagrams

SANDI SETIAWAN

Life Member, Clare Hall, University of Cambridge

PREFACE

Why We Look Beyond the Event Horizon

> *"Black holes are not eternal prisons as we once thought. They are sources and sinks of energy, and possibly gateways to new understanding."*
> — *Stephen Hawking*

This book is a journey—across a century of discovery, from the first stirrings of Einstein's gravity to the quantum whispers of Hawking radiation, and far beyond. Black holes, once mere solutions to equations, have become cosmic laboratories that challenge our deepest understanding of space, time, energy, and information.

What You Will Find In This Book

- The discovery and nature of black holes
- The geometry of spacetime and the event horizon
- Hawking radiation and the fate of black holes
- The information paradox and modern resolutions
- Classical and quantum energy extraction mechanisms
- Astrophysical jets, accretion disks, and observational evidence
- Future technologies and the implications for humanity

The Purpose of this Book

Our purpose is twofold:

- To explain black holes and their evolution, from their birth in collapsing stars to the remarkable phenomenon of Hawking radiation.
- To explore how energy and information may be extracted—through Hawking radiation and other mechanisms—opening new possibilities for science, technology, and our understanding of the universe.

We weave together classical general relativity, quantum field theory, and modern developments in quantum information and holography to present a unified picture that is both accessible and forward-looking.

For Whom This Book Is Written

This book is written for advanced students, researchers, and curious minds who wish to see the big picture—while also diving into the details. No prior knowledge beyond undergraduate physics is assumed, but an open mind is essential.

Whether you are a physicist, an engineer, or simply a dreamer, you are most welcome.

A Story of Unity

From Newton's gravity to Einstein's spacetime, from Hawking's quantum insight to today's holographic frontiers, the story of black holes is a story of human curiosity refusing to stop at the horizon.

This book invites you to step beyond that horizon with us.

Our Roadmap

The Foundations of Gravity → The Birth and Life of Black Holes → Horizons and the Fabric of Spacetime → Quantum Effects and Hawking Radiation → Information, Paradox, and Resolution → Energy Extraction Mechanisms → Applications and Future Horizons

An Invitation

Black holes are mystery and opportunity. They are the universe's ultimate engines and its deepest puzzles. As we advance, we may learn not only how to understand them—but perhaps how to work with them.

Let us begin this journey—
beyond the event horizon.

The universe is vast. Our curiosity is boundless. The horizon is only the beginning.

Table of Contents

Beyond the Causal Horizon

Hawking Radiation, Other Energy Extractions from Black Holes, and Penrose Diagrams

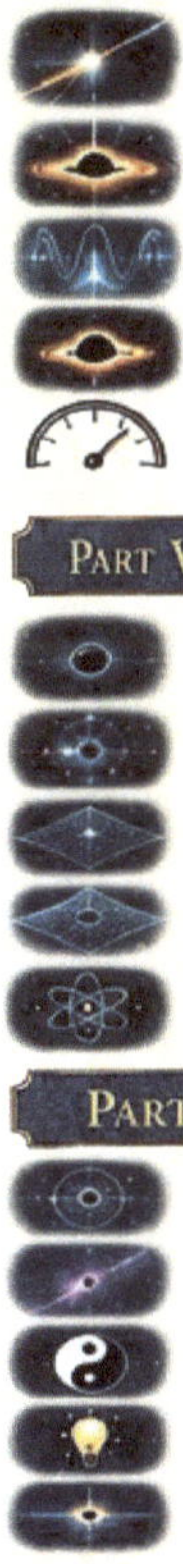

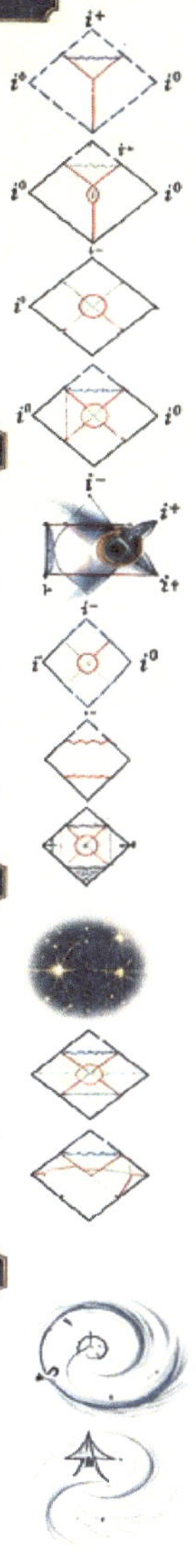

Part I — Foundations of Black Holes

Part II — Singularity and Rotation

Part III — Hawking Radiation

Part IV — Evaporation and Extraction

Part V — Wave and Field Extraction

Part VI — Energy and Quantum Fields

Part VII — Information and Quantum Gravity

Part VIII — Future Frontiers

Symbol & Line Conventions

- Null infinity (boundary)
- Curvature singularity ($r = 0$)
- Event horizon / Apparent horizon
- Timelike infinity (i^+, i^-)
- Null surface (constant r or v)

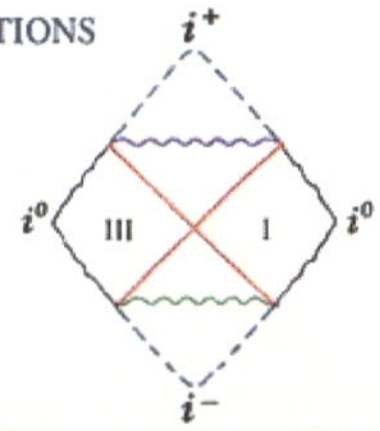

Region Labels (typical)

- I : Exterior region (observable universe)
- II : Other asymptotic region / exterior
- III : Interior region (inside horizon)
- IV : Black hole interior / trapped region
- V, VI, ... : Additional regions by extension

Journey Map

“Black holes are not the end of everything. They are the beginning of new physics and new possibilities.”

— Stephen Hawking

1

Beyond the Event Horizon

From Hawking Radiation to Other Energy Extractions from Black Holes

The Birth of Black Holes

Black holes are among the most fascinating predictions of Einstein's theory of general relativity. They are born when massive stars collapse under their own gravity, squeezing matter into an incredibly small region where space and time are warped beyond imagination.

Inside lies a singularity—where our current laws of physics break down—and surrounding it, the event horizon, the point of no return. Yet black holes are not completely dark. Thanks to quantum effects, they can radiate energy, interact with their surroundings, and power some of the most energetic phenomena in the universe.

A Journey of Discovery

Einstein 1915	Schwarzschild 1916	Penrose 1965	Hawking 1974
General relativity predicts black holes.	First exact solution reveals the event horizon.	Singularity theorems prove black holes naturally form.	Quantum effects allow black holes to radiate.

What is a Black Hole?

A black hole is a region of spacetime where gravity is so strong that nothing—not even light—can escape once it crosses the event horizon.

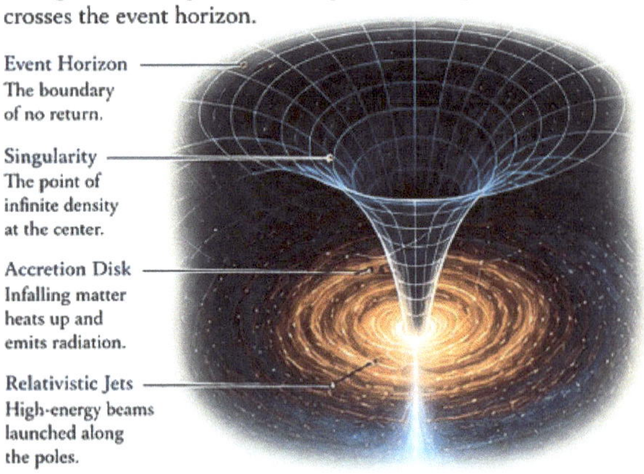

Hawking Radiation — Light from the Void

In 1974, Stephen Hawking showed that quantum effects near the event horizon can create particle–antiparticle pairs. One falls in, while the other escapes—carrying energy away. Over vast timescales, this causes the black hole to lose mass and eventually evaporate.

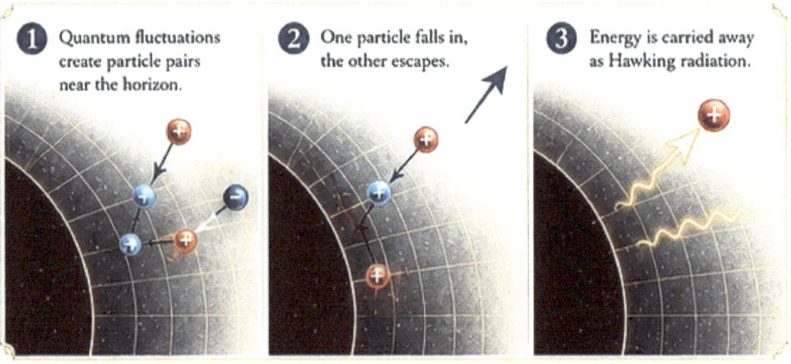

Why it Matters

Black holes power quasars, active galactic nuclei, and gamma-ray bursts.

They connect gravitational physics with quantum mechanics.

They challenge our understanding of time, information, and the ultimate fate of matter.

Observations from across the spectrum reveal their hidden presence.

Studying black holes may unlock new sources of energy for the future.

"To understand the universe, we must dare to go beyond the event horizon."

2

Gravity Through History

THE LONG ROAD TO BLACK HOLES

Our understanding of gravity has evolved over centuries.
Each new idea brought us closer to the concept of black holes.

Isaac Newton
1687

Published the law of universal gravitation. Gravity is a force acting at a distance.

$$F = G\frac{m_1 m_2}{r^2}$$

Albert Einstein
1915

Developed general relativity. Gravity is not a force, but the curvature of spacetime.

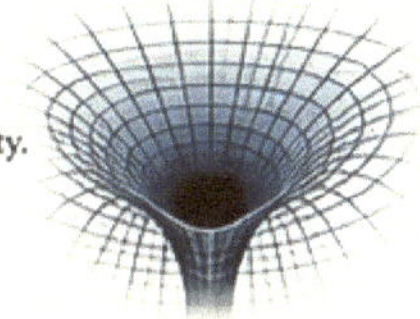

Karl Schwarzschild
1916

Found the first exact solution to Einstein's equations for a spherical mass. Hidden within it lies an event horizon.

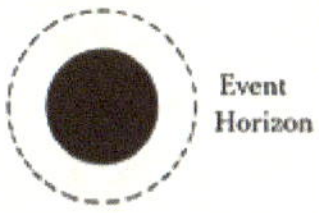

Schwarzschild radius $r_s = \frac{2GM}{c^2}$

S. Chandrasekhar
1930s

Showed that massive stars can collapse under gravity if their mass is above a certain limit.

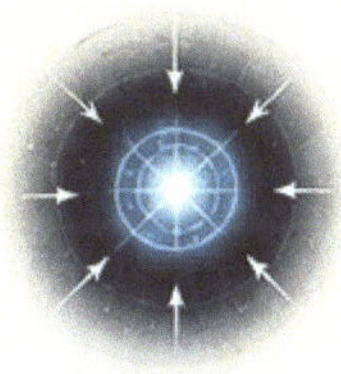

J. Robert Oppenheimer
1939

Along with Snyder, proposed gravitational collapse can continue to a singularity.

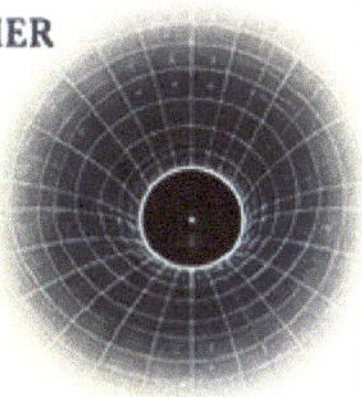

Spacetime Curvature

In general relativity, mass and energy tell spacetime how to curve, and curved spacetime tells matter how to move.

"Spacetime tells matter where to go; matter tells spacetime how to curve."
— John Archibald Wheeler

From Stars to Black Holes

Stellar Birth	Stellar Life	Core Collapse	Black Hole
	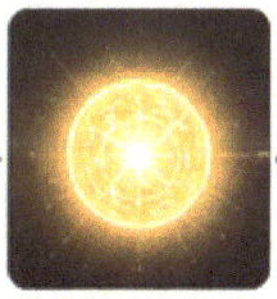		
Gas and dust collapse to form a star.	Nuclear fusion shines for millions to billions of years.	Fuel runs out. The core collapses under gravity.	If the core is massive enough, nothing can stop the collapse.

"The most incomprehensible thing about the universe is that it is comprehensible."
— Albert Einstein

Key Milestones on the Path to Black Holes

1687
Newton's law of gravitation

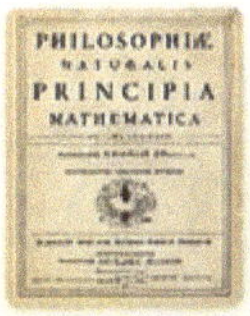

1915
Einstein's general relativity

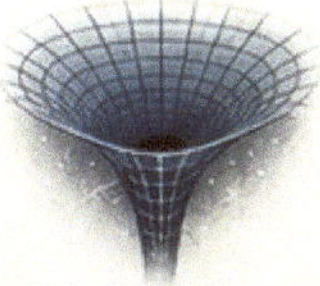

1916
Schwarzschild solution

1930s
Chandrasekhar limit

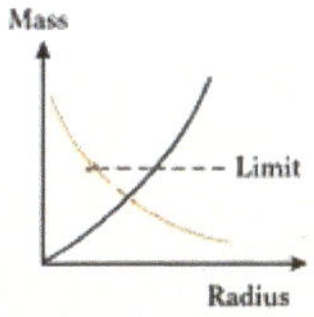

1939
Oppenheimer–Snyder collapse model

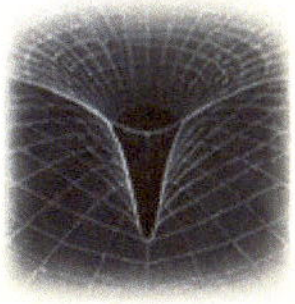

1965
Penrose singularity theorems

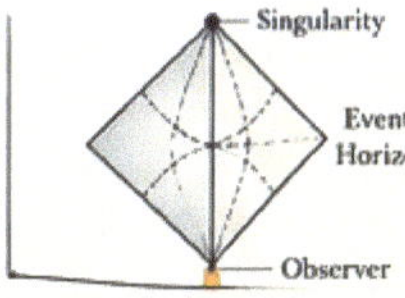

1974
Hawking radiation discovery

The Big Picture

From Newton's invisible force to Einstein's curved spacetime, and from stellar collapse to quantum radiation—black holes sit at the crossroads of gravity, quantum mechanics, and the structure of the cosmos.

They are not just the end of stars, but the beginning of new physics.

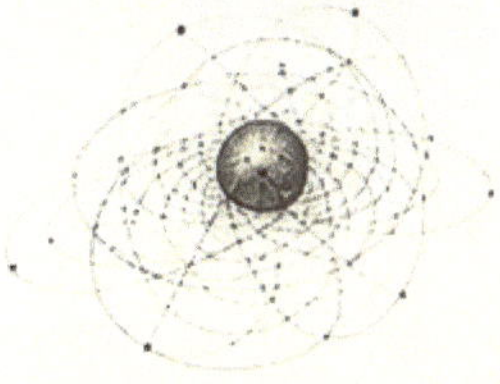

3 Birth of a Black Hole

When Stars Die, Black Holes Are Born

Black holes are born from the gravitational collapse of massive stars. When a star exhausts its nuclear fuel, gravity takes over.

The Life and Death of a Massive Star

1 Main Sequence Star
Hydrogen fusion in the core produces energy, balancing gravity. The star shines steadily for millions to billions of years.

2 Red Supergiant
Fuel runs low. The core contracts, the outer layers expand and cool. The star becomes unstable.

3 Core Collapse
The iron core cannot produce energy. Gravity overwhelms it, and it collapses in a fraction of a second.

4 Supernova Explosion
The outer layers are blown away in a tremendous explosion. Heavy elements are scattered into space.

5 Black Hole
The collapsed core, now extremely dense, becomes a black hole. Its gravity prevents anything from escaping.

Gravity's Inevitable Victory

When the core of a massive star exceeds about 3 times the mass of our Sun, no known force can stop it from collapsing. Electrons and protons combine to form neutrons, but even neutron pressure cannot hold up the core. It collapses to a point so dense that spacetime itself is warped beyond escape.

"Nature hides nothing; it is we who are not yet ready to understand."
— Albert Einstein

The Stages of Collapse

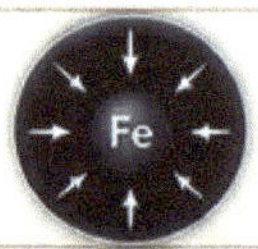

1 Iron Core Forms
Fusion builds heavier elements until iron forms. Iron cannot release energy through fusion.

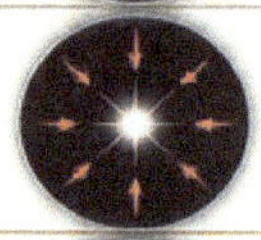

2 Instability
The core becomes unstable and begins to collapse under its own gravity.

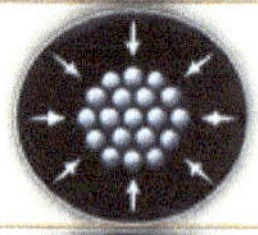

3 Neutronization
Electrons are forced into protons, forming neutrons. The core becomes a neutron star briefly.

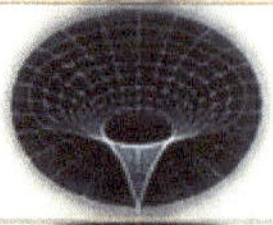

4 Black Hole Formation
If the core is too massive, the neutron star collapses further into a singularity, forming a black hole.

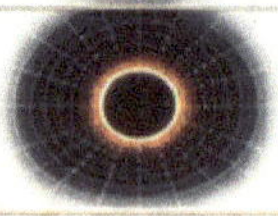

5 Event Horizon
A boundary forms beyond which nothing—not even light—can escape.

Key Facts About Stellar Black Holes

Mass Range
Typically formed from stars with 20–100+ times the mass of the Sun.

Size
A stellar black hole has a radius of only a few tens of kilometers.

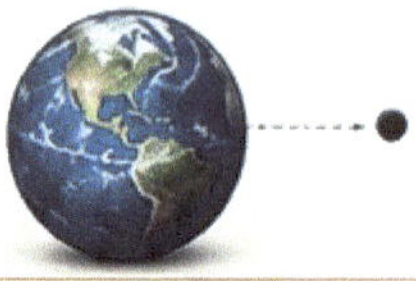

Density
More dense than an atomic nucleus—mass of a mountain in a sugar-cube sized volume.

No Escape
Escape velocity exceeds the speed of light at the event horizon.

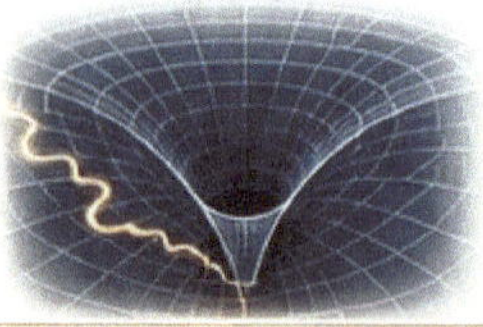

Element Factory
Supernovae create and spread heavy elements essential for planets and life.

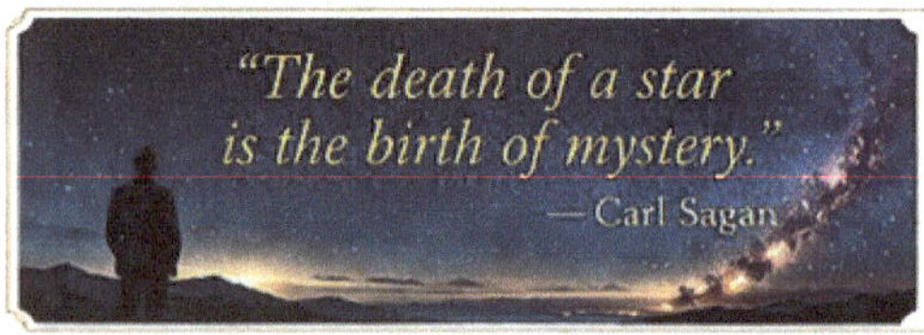

The Takeaway

Black holes are not accidents—they are natural outcomes of the universe. They shape galaxies, influence stars, and may hold the keys to the deepest questions in physics.

The universe recycles the stars—and from their ashes, black holes are born.

4

Inside Schwarzschild Geometry

The Curved Stage Where Time and Space Change Roles

Karl Schwarzschild solved Einstein's equations in 1916 and discovered a remarkable solution that describes the spacetime outside a spherical, non-rotating mass. Hidden within it lies the interior of a black hole.

Karl Schwarzschild

1873–1916

A brilliant German physicist and soldier, Schwarzschild found the first exact solution to Einstein's field equations for the spacetime around a spherical mass.

"The solution was too strange to be a mere mathematical curiosity."

— Schwarzschild

The Schwarzschild Metric

The spacetime outside a spherical mass M is described by:

$$ds^2 = -\left(1 - \frac{2GM}{rc^2}\right)c^2dt^2 + \left(1 - \frac{2GM}{rc^2}\right)^{-1}dr^2 + r^2(d\theta^2 + \sin^2\theta\, d\phi^2)$$

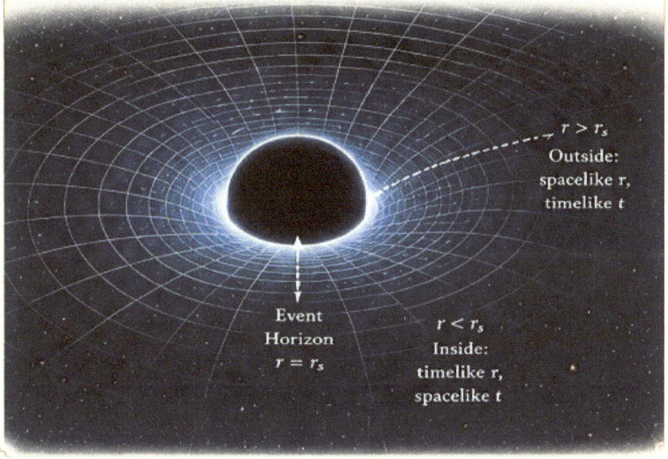

Key Insight

At the event horizon

$$r_s = \frac{2GM}{c^2},$$

the nature of space and time interchange. Inside the horizon, all paths lead inward.

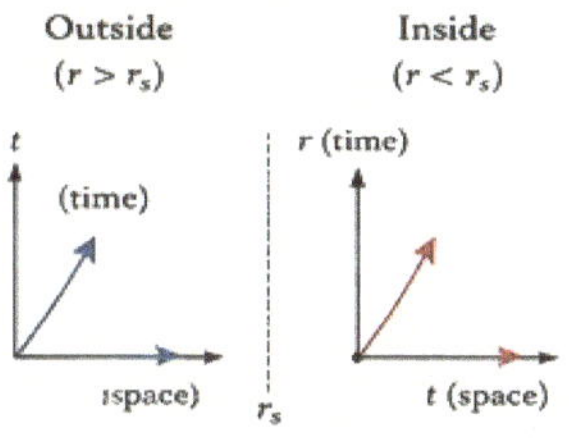

You cannot stand still.
You cannot turn back.
You can only move toward smaller r.

The Event Horizon

The radius $r_s = \frac{2GM}{c^2}$ is the boundary beyond which escape is impossible. Not even light can get out.

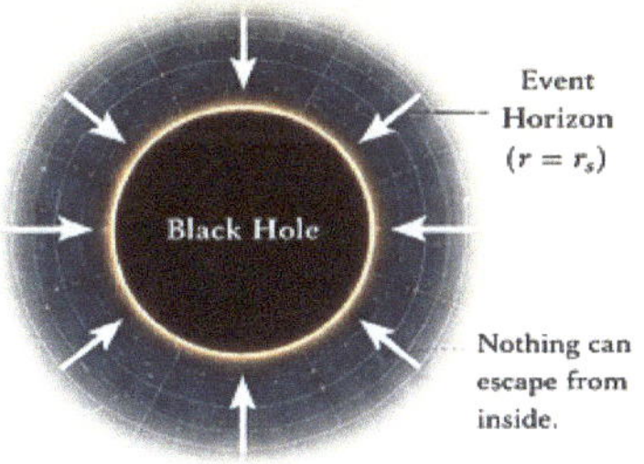

Geodesics: Paths in Curved Spacetime

Inside the horizon, all future-directed paths (geodesics) head toward $r = 0$, the singularity.

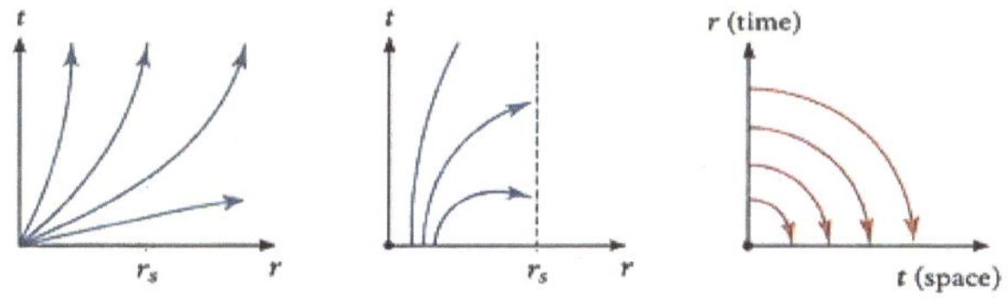

Outside ($r > r_s$)	At Horizon ($r = r_s$)	Inside ($r < r_s$)
Particles can move away or toward the mass.	Light takes an infinite time to reach the horizon.	All paths inevitably lead to the singularity at $r = 0$.

The Spacetime Diagram

A Penrose diagram shows the global structure of the Schwarzschild black hole.

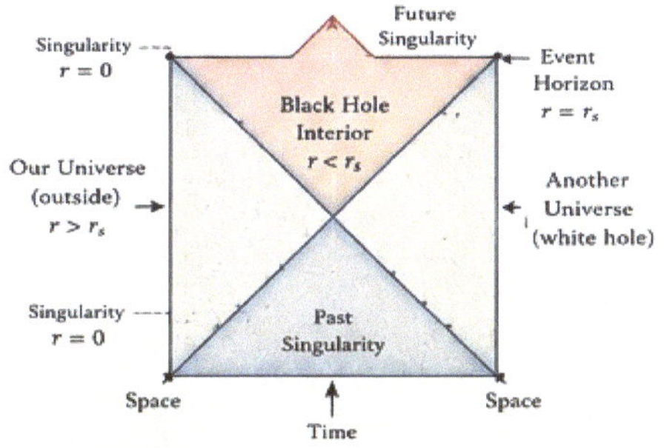

What an Observer Sees

An outside observer never sees anything cross the horizon. To them, infalling matter appears to freeze and fade.

1. **Far away** Everything looks normal.
2. **Approaching** The object appears dim and redshifted.
3. **Near horizon** It slows down and reddens.
4. **At horizon** It seems to freeze and fade.
5. **Gone** It never crosses in our view.

Key Lessons

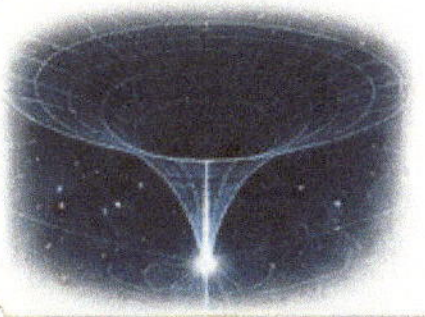

- Black holes are solutions of Einstein's equations, not science fiction.
- At the horizon, the rules of space and time profoundly change.
- Inside, nothing can escape; the singularity is unavoidable.
- Understanding this geometry is the key to modern black hole physics.

The Singularity

At $r = 0$, curvature becomes infinite. Our current theories break down. This is where quantum gravity must begin.

"The Schwarzschild solution revealed that general relativity predicts not just curved spacetime, but a one-way gate from which nothing returns."

5

EVENT HORIZON

THE POINT OF NO RETURN

The event horizon is the boundary beyond which nothing—
not even light—can escape a black hole.
It marks the ultimate limit set by gravity.

SPACETIME NEAR A BLACK HOLE

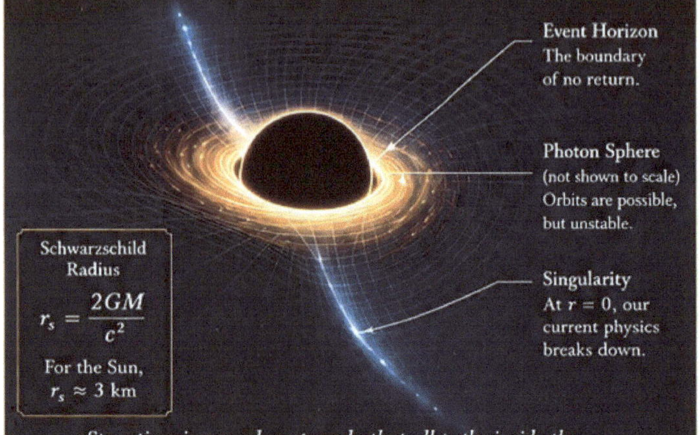

Schwarzschild Radius

$$r_s = \frac{2GM}{c^2}$$

For the Sun, $r_s \approx 3$ km

Spacetime is curved so strongly that all paths inside the event horizon lead inevitably to the singularity.

WHAT IS THE EVENT HORIZON?

Imagine standing outside a black hole.
If you cross a certain distance r_s, escape becomes impossible.
No signal, no matter how fast, can reach the outside universe.

OUTSIDE ($r > r_s$)
Escape is possible.

INSIDE ($r < r_s$)
Escape is impossible.

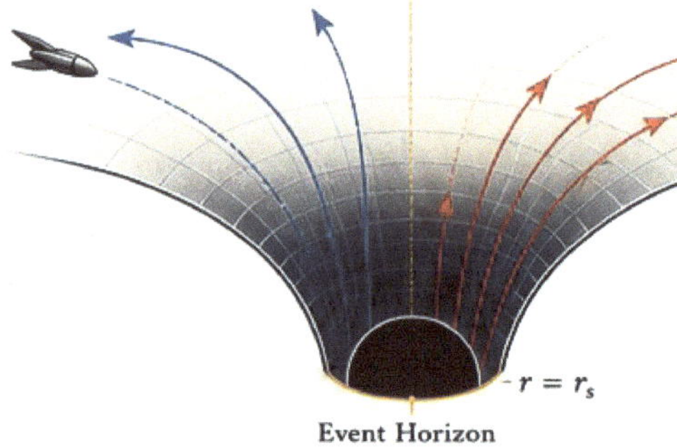

Every possible path bends inward.

Time itself carries you closer to the singularity.

LIGHT CONE VIEW

The light cone shows the possible directions light can travel. At the event horizon, the cone tips inward.

1 FAR AWAY ($r > r_s$)

The light cone opens upward and downward in time.

Future and past are both accessible.

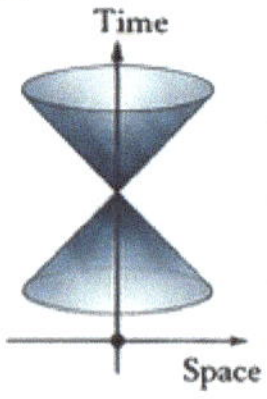

All directions possible.

2 AT THE HORIZON ($r = r_s$)

The cone tilts.

The outward direction lies along the horizon.

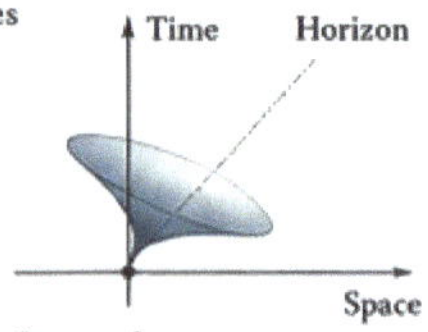

Outward escape just becomes impossible.

3 INSIDE ($r < r_s$)

The cone tips fully inward.

The future points toward the singularity.

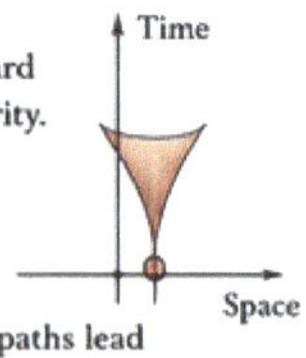

All paths lead inward.

ONE-WAY MEMBRANE

To an outside observer, the event horizon behaves like a one-way surface.

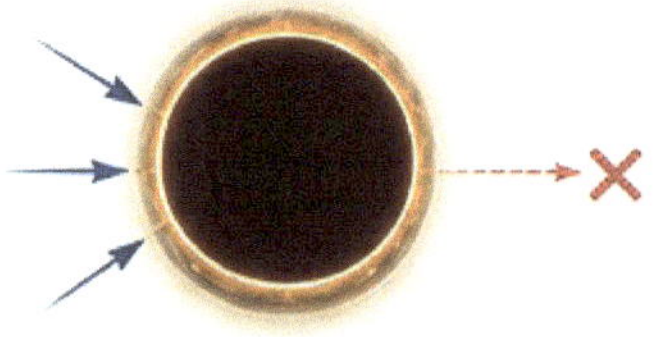

Inward: allowed
Outward: forbidden

THE HORIZON IN DIFFERENT BLACK HOLES

STELLAR BLACK HOLE

$$r_s = \frac{2GM}{c^2}$$

A few kilometers.

SUPERMASSIVE BLACK HOLE

$$r_s = \frac{2GM}{c^2}$$

Millions to billions of kilometers.

ROTATING (KERR) BLACK HOLE

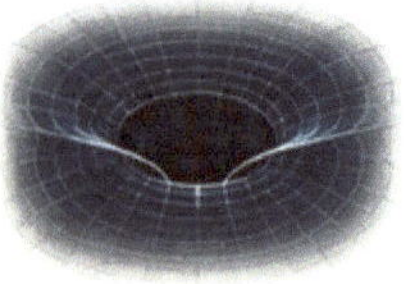

Two horizons:
Outer r_+ and inner r_-.
Depend on spin a.

CHARGED (REISSNER–NORDSTRÖM)

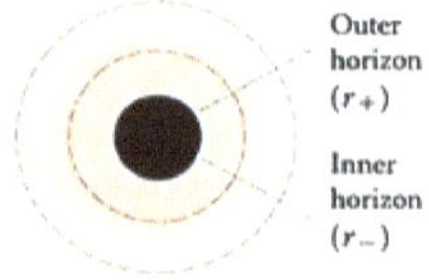

Two horizons if $Q^2 < GM^2$.
Extremal case: $r_+ = r_-$.

IN FALLING FRAME

For someone crossing the horizon, nothing unusual is felt locally.
The horizon is not a wall—
it is a feature of spacetime itself.

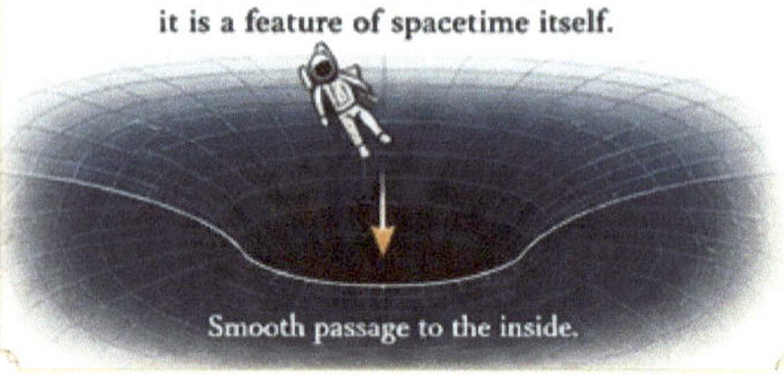

KEY POINTS

- The event horizon is defined globally, not locally.
- It depends on the entire future of spacetime.
- Crossing it is not an explosion— it is a change of causal structure.
- Information that falls in cannot influence the outside universe.

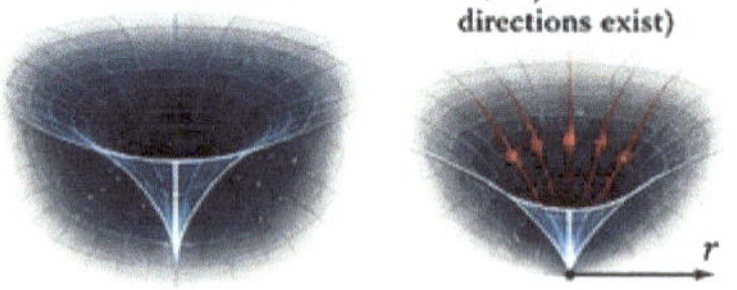

MATHEMATICAL DEFINITION

The event horizon is the boundary of the causal past of future null infinity ($\mathscr{I}^+$).

$$\mathcal{H}^+ = \partial J^-(\mathscr{I}^+)$$

It is a null hypersurface generated by light rays that never reach $\mathscr{I}^+$.

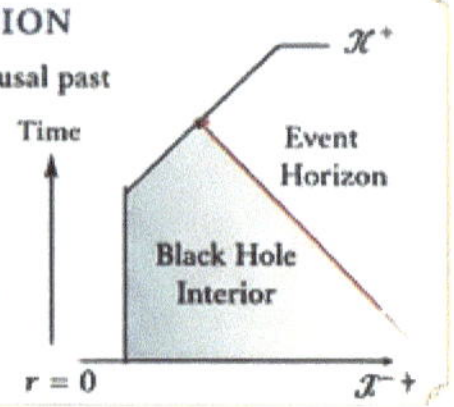

WHY IT MATTERS

Defines the boundary of black holes.

Shapes the growth of black holes and galaxies.

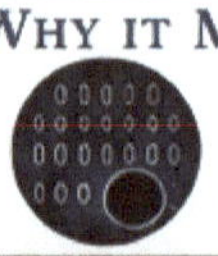

Central to information, quantum gravity, and holography.

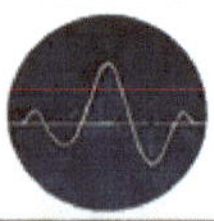

A natural laboratory for extreme physics and new ideas.

Challenges our deepest understanding of space, time, and reality.

"The event horizon is not the end of space, but the beginning of profound new mysteries."

6

SINGULARITY

THE LIMIT WHERE PHYSICS BREAKS DOWN

At the center of a black hole lies a singularity—where curvature becomes infinite and the known laws of physics cease to work. It is a boundary of our current theories and a gateway to new physics.

WHAT IS A SINGULARITY?

A singularity is a place where matter is crushed to zero volume and density becomes infinite. Space and time as we know them end.

It is not an object, but a prediction that our current theories are incomplete.

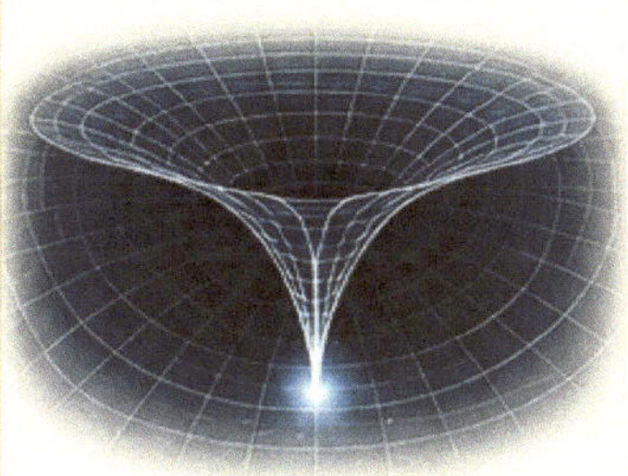

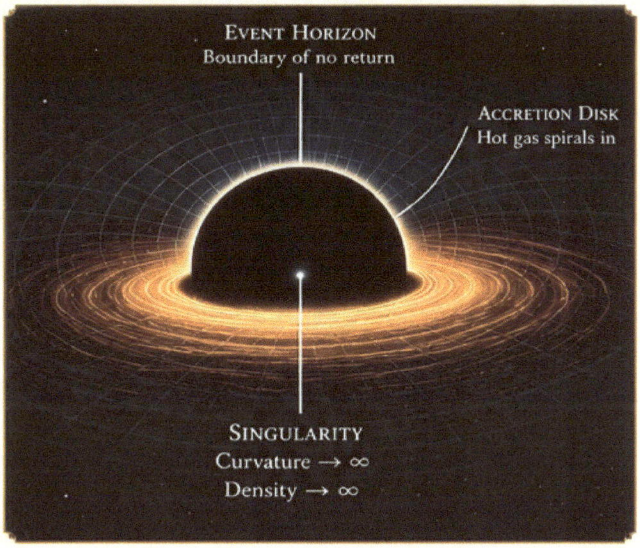

CURVATURE BLOW-UP

General relativity tells us that as we approach the singularity ($r \to 0$), the curvature of spacetime grows without bound.

Kretschmann scalar:

$$K = R_{abcd}R^{abcd} = \frac{48G^2M^2}{c^4 r^6}$$

$$\text{As } r \to 0, \quad K \to \infty$$

No physical object can survive this environment.

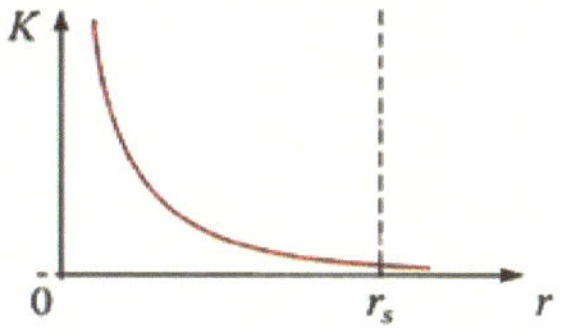

INSIDE THE EVENT HORIZON

Once inside the event horizon, all paths in spacetime are forced toward smaller r.

The singularity is inevitable.

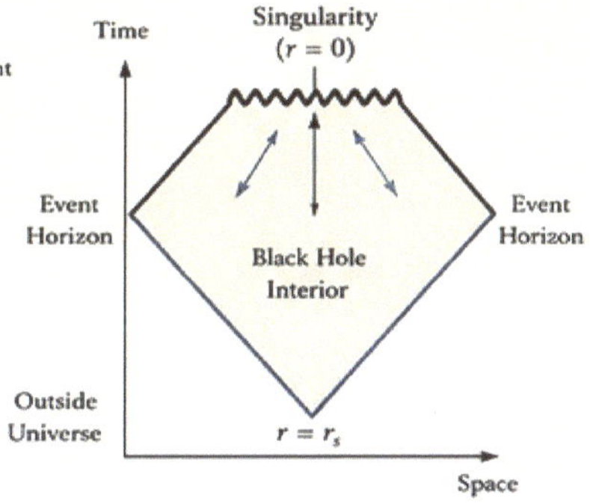

WHAT HAPPENS AT THE SINGULARITY?

1 Infinite Density

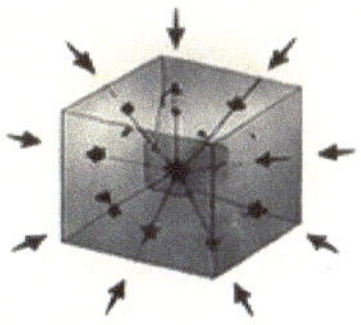

Matter is crushed into zero volume.

2 Infinite Curvature

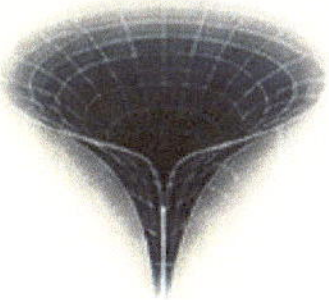

Spacetime curvature becomes infinite.

3 End of Predictability

The laws of physics as we know them cease to predict.

4 Unknown Physics

Quantum gravity is needed to go beyond.

NOT A PHYSICAL OBJECT

The singularity is a feature of the mathematical solution, not something we can observe directly.

It signals a breakdown of general relativity, just as infinity in Newtonian gravity signaled a need for new ideas.

"The singularities show that our present theories must be modified."

— Roger Penrose

EXAMPLES OF SINGULARITIES

- Schwarzschild black hole: point singularity at $r = 0$
- Kerr black hole: ring singularity
- Reissner–Nordström: still a central singularity
- Cosmology (Big Bang): the universe began in a singular state?

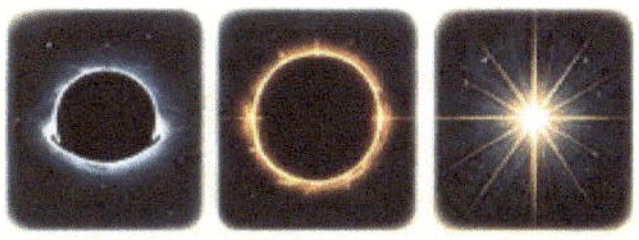

THE NEED FOR NEW THEORY

To understand what lies beyond the singularity, we need a theory of quantum gravity that unifies:

- Quantum mechanics
- General relativity
- Information theory
- Geometry of spacetime

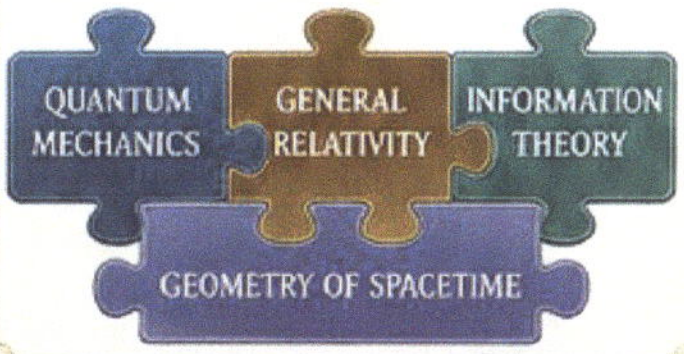

POSSIBLE PATHS BEYOND

Many ideas attempt to resolve or replace singularities:

- Loop quantum gravity
- String theory
- Asymptotic safety
- Holographic principle
- Non-singular black holes

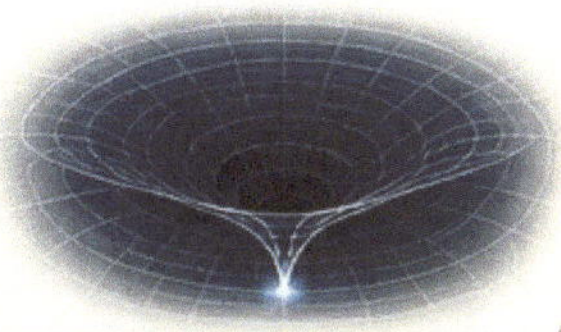

KEY TAKEAWAYS

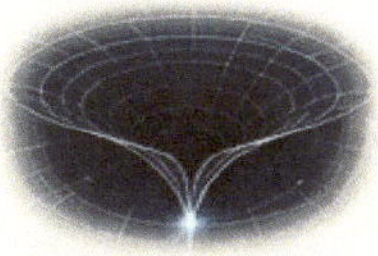

Singularity is where curvature and density become infinite.

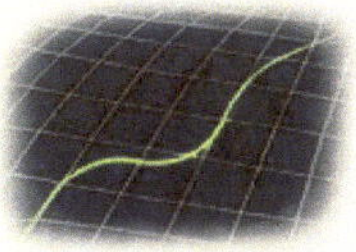

Inside the horizon, all paths lead to the singularity.

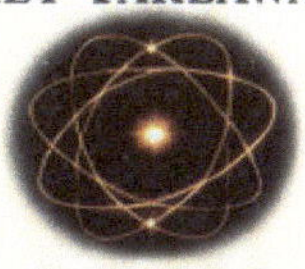

Current physics breaks down—new theory is required.

Singularity may hide answers to the deepest mysteries of nature.

Understanding it may unlock the future of physics and cosmology.

"The singularity is not where the story ends, but where our understanding is challenged to begin."

7 Different Families of Black Holes

Not All Black Holes Are the Same

Black holes come in various sizes and forms.
They are shaped by the mass of their progenitor stars and their cosmic environments.

1. Stellar-Mass Black Holes

Mass: ~ 3 – 100 $M_\odot$

Formed when a massive star collapses at the end of its life.

- Found in binary star systems (can emit X-rays as they accrete matter
- May merge, producing gravitational waves

These are the most common black holes in our galaxy.

2. Intermediate-Mass Black Holes

Mass: ~ 100 – 100,000 $M_\odot$

Possible "missing link" between stellar and supermassive black holes.

- Found in dense star clusters?
- Evidence is still uncertain
- May form from runaway stellar collisions

Their existence is one of the great mysteries today.

3. Supermassive Black Holes

Mass: ~ 10^6 – 10^{10} $M_\odot$

Reside at the centers of most galaxies, including our Milky Way.

- Power quasars and active galactic nuclei
- Influence galaxy evolution
- Extremely luminous when active

Sagittarius A* is the supermassive black hole at our galaxy's center.

4. Primordial Black Holes

Mass: could range from tiny (10^{15} g) to very massive

Hypothesized to form in the early Universe, not from stars.

- Could be dark matter candidates
- Might explain some cosmic ray events
- Still purely theoretical

If they exist, they could hold secrets of the early Universe.

The Mass Scale of Black Holes

10^{15}g — 10^{20}g — 10^{25}g — 10^{30}g — 10^{35}g — 10^{40}g

Primordial (Tiny)

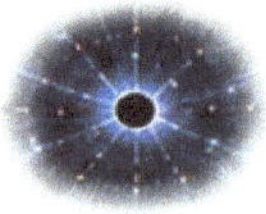

Stellar-Mass (3 – 100 $M_\odot$)

Intermediate-Mass (100 – 100,000 $M_\odot$)

Supermassive (10^6 – 10^{10} $M_\odot$)

Beyond Known (Theoretical)

How Black Holes are Formed

1. Nebula

A giant cloud of gas and dust in space.

2. Star Formation

Gravity pulls matter together to form stars.

3. Massive Star

Massive stars live fast and burn their fuel quickly.

4. Core Collapse

When fuel runs out, the core collapses.

5. Supernova

The star explodes, scattering its outer layers.

6. Black Hole

The core collapses further into a black hole.

Observing Black Holes

- X-rays: from hot gas near the black hole
- Radio waves: from jets and accretion disks
- Star orbits: stars reveal hidden giants (e.g., Sgr A*)
- Gravitational waves: from black hole collisions

Where Do We Find Them?

In binary star systems in our galaxy

At the centers of galaxies

In dense star clusters and globular clusters

Possibly throughout the early Universe

Why It Matters

- Black holes test the limits of general relativity.
- They influence galaxy formation and cosmic evolution.
- They may unlock the mystery of dark matter and quantum gravity.
- They inspire new technologies and new ways of thinking.

"From the tiniest whispers of the early Universe to the giants at the hearts of galaxies, black holes come in many forms—yet all share the same defining mystery: the power of gravity without escape."

8

The Kerr Black Hole

Rotation Warps Spacetime

Most astrophysical black holes spin. In 1963, Roy Kerr discovered a solution describing rotating black holes. Rotation drags spacetime itself, creating an ergosphere—a region where energy can be extracted.

Roy Kerr

1934–2003

New Zealand mathematician who found the rotating black hole solution to Einstein's equations.

"Einstein's equations still had a few surprises left for us."

— Roy Kerr

Spacetime Around a Rotating Black Hole

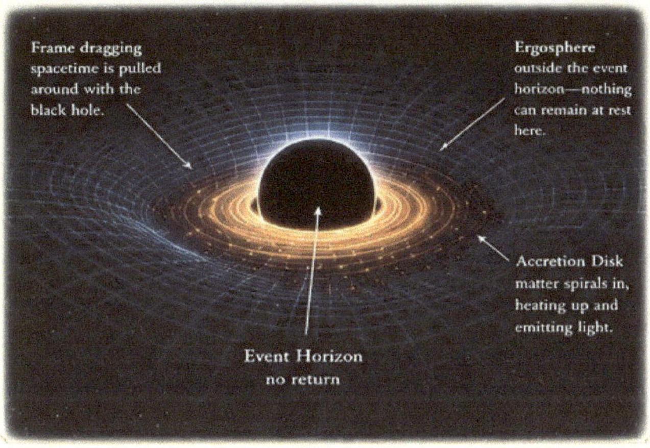

The Kerr Metric (Boyer–Lindquist Form)

The spacetime outside a rotating mass M with angular momentum $J = aM$ is given by the Kerr metric:

$$ds^2 = -\left(1 - \frac{2Mr}{\Sigma}\right) dt^2 - \frac{4Mar\sin^2\theta}{\Sigma}\, dt\, d\phi + \frac{\Sigma}{\Delta} dr^2 + \Sigma\, d\theta^2 + \left(r^2 + a^2 + \frac{2Ma^2 r \sin^2\theta}{\Sigma}\right)\sin^2\theta\, d\phi^2$$

where $\Sigma = r^2 + a^2\cos^2\theta,$

$\Delta = r^2 - 2Mr + a^2$

Key Parameters

- M : mass of the black hole
- a : spin parameter ($0 \le a \le M$)
- $a = 0$: Schwarzschild black hole
- $a = M$: maximally rotating black hole

Horizons and Ergosphere

Event Horizons

Kerr black holes have two horizons (when $a < M$):

$$r_\pm = M \pm \sqrt{M^2 - a^2}$$

r_+ : outer event horizon (black hole boundary)

r_- : inner (Cauchy) horizon

When $a = M$, the horizons coincide:

$r_+ = r_- = M$ (extremal Kerr black hole)

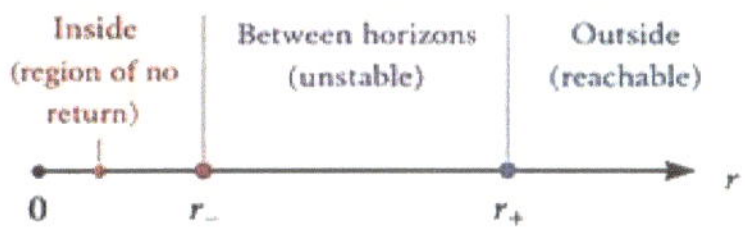

The Ergosphere

The ergosphere is the outer boundary where no object can remain at rest.

$$r_{erg} = M + \sqrt{M^2 - a^2\cos^2\theta}$$

At the equator ($\theta = \pi/2$): $r_{erg} = 2M$

At the poles ($\theta = 0, \pi$): $r_{erg} = r_+$

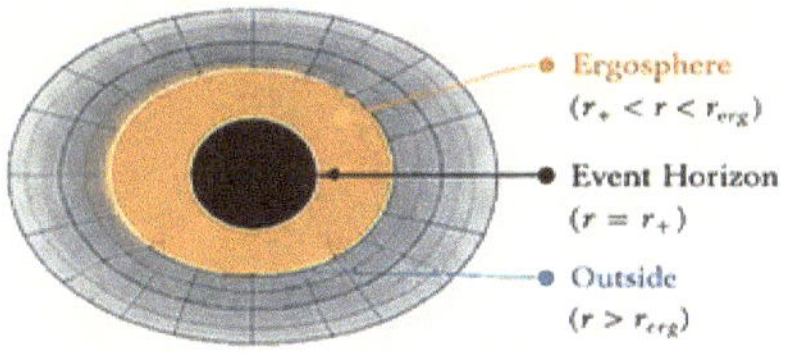

The Ergosphere in 3D

The ergosphere is oblate—flattened at the poles and bulging at the equator.

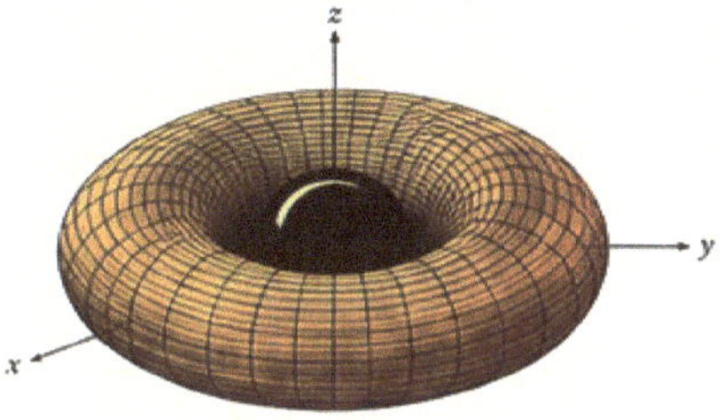

Frame Dragging

Inside the ergosphere, spacetime is dragged around the black hole. All particles are forced to co-rotate in the direction of spin.

Frame dragging becomes stronger closer to the black hole and faster near the equator.

The Penrose Process (Energy Extraction)

Inside the ergosphere, a particle can split into two. One falls into the black hole with negative energy, the other escapes with more energy than the original.

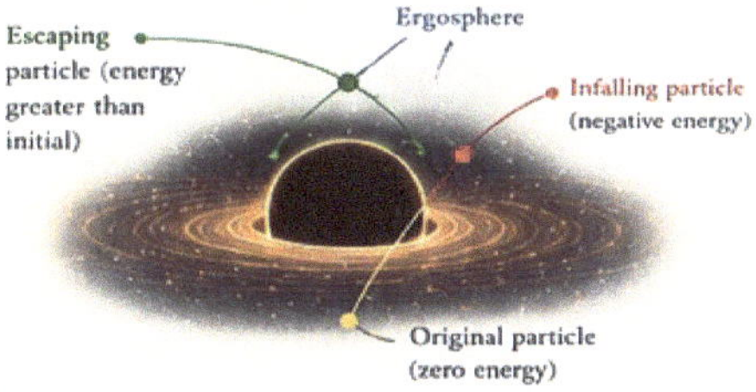

Superradiance

Waves (like light) can also extract energy from a rotating black hole.

Occurs when: $0 < \omega < m\,\Omega_H$

where

ω = wave frequency,

m = azimuthal quantum number,

$\Omega_H = \dfrac{a}{2Mr_+}$ = angular velocity of the horizon.

The reflected wave is amplified!

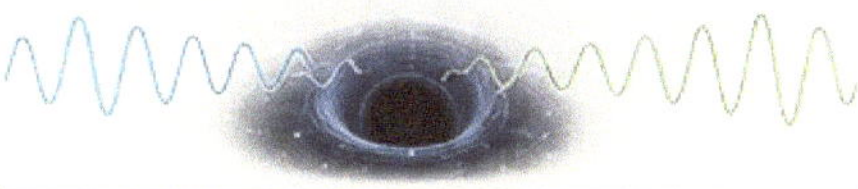

Types of Kerr Black Holes

Slowly Rotating ($a \ll M$)
Close to Schwarzschild.
Small frame dragging.

Moderately Rotating ($0 < a < M$)
Distinct ergosphere and two horizons.
Strong astrophysical relevance.

Extremal ($a = M$)
Horizons coincide. Maximum spin.
Unique theoretical properties.

Key Takeaways

- Rotation changes everything. It drags spacetime and creates an ergosphere.
- The ergosphere is the engine room for energy extraction.
- Kerr black holes power jets, quasars, GRBs, and some of the most energetic events in the Universe.
- They are laboratories for testing general relativity in the strong-field regime.

Astrophysical Evidence

Many observations indicate black holes are rotating.

Relativistic jets in quasars and AGNs require spin.

X-ray reflection and iron line profiles reveal fast rotation.

Gravitational waves from mergers measure final spin.

Accretion disk modeling shows high spin in many systems.

"Rotation gives a black hole not just gravity, but the power to shape energy, light—and the cosmos itself."

9

Frame Dragging

Spacetime is Dragged Around a Rotating Black Hole

According to general relativity, rotating mass does not only curve spacetime—
it drags spacetime itself in the direction of rotation.
This effect is called frame dragging, or the Lense–Thirring effect.

The Basic Idea

Imagine stirring honey with a spoon. The honey swirls around the spoon. Similarly, a rotating black hole "stirs" spacetime, causing nearby objects and even light to swirl in the same direction.

Spacetime Dragged by a Rotating Black Hole

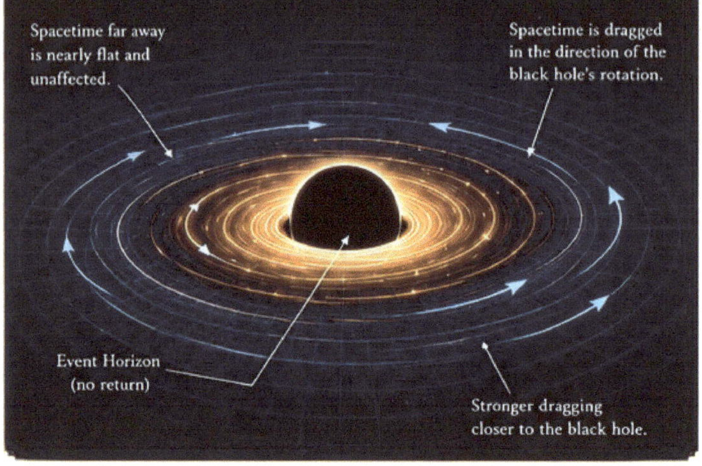

The Lense–Thirring Effect

First predicted independently by Joseph Lense and Hans Thirring in 1918.

A rotating mass with angular momentum J produces a "gravitomagnetic" field $\vec{B}_g$ that causes nearby gyroscopes to precess.

$$\Omega_{LT} = \frac{2G}{c^2 r^3} J$$

where

- G = gravitational constant
- c = speed of light
- r = distance from the mass
- Ω_{LT} = precession (dragging) angular velocity

How Frame Dragging Works

A gyroscope orbiting a rotating black hole will gradually turn in the direction of the black hole's spin.

Without frame dragging (spacetime not rotating) | With frame dragging (spacetime rotating)

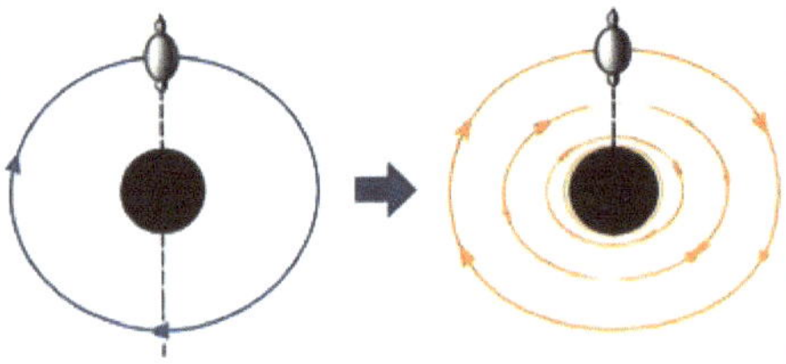

Gyroscope maintains its orientation. | Gyroscope's axis slowly precesses in the direction of rotation.

Mathematical Description (Intuition)

In the Kerr metric, off-diagonal terms $g_{t\phi}$ appear, coupling time and angular coordinates.

$$ds^2 = g_{tt}dt^2 + 2g_{t\phi}dt\,d\phi + g_{\phi\phi}d\phi^2 + \cdots$$

The term $g_{t\phi} \neq 0$ is what causes the dragging of inertial frames.

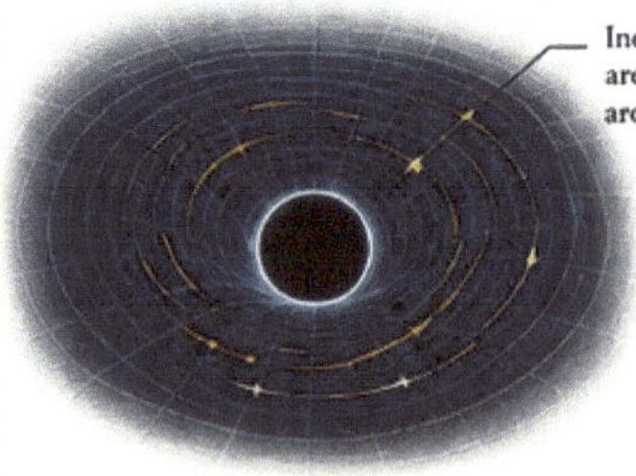

Strength of Frame Dragging

Stronger closer to the black hole and near the equatorial plane.

Dependence (roughly):

$$|\Omega_{LT}| \propto \frac{aM}{r^3}$$

where

- M = mass of black hole
- $a = J/M$ = spin parameter
- $0 \leq a \leq M$
- r = distance from black hole

Maximum in the equatorial plane, decreasing with distance.

Measuring Frame Dragging: Gravity Probe B

NASA's Gravity Probe B (launched 2004) measured the precession of gyroscopes around Earth caused by frame dragging.

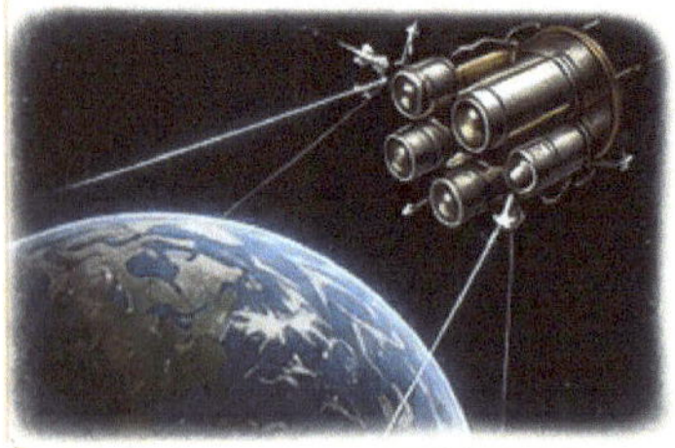

Measured precession 39.2 ± 7.2 milliarcseconds per year.

Result agrees with Einstein's prediction within experimental uncertainty!

Astrophysical Consequences

Accretion Disks

Inner parts of the disk are dragged, affecting jet launching and radiation.

Jet Alignment

Jets from AGNs and quasars may be aligned by frame dragging near the hole.

Orbital Precession

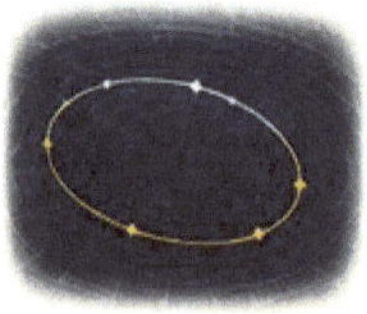

Orbits of stars and gas near the black hole can precess due to dragging.

Gravitational Waves

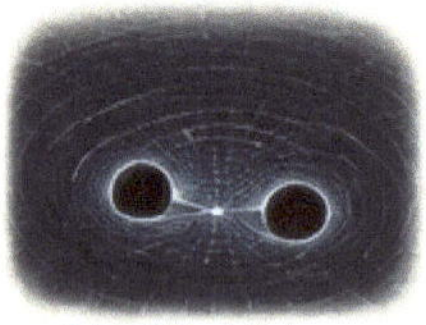

Spin of black holes affects gravitational wave signals from mergers.

Key Takeaways

- Rotating black holes drag spacetime around with them.
- Frame dragging is a real, measurable prediction of general relativity.
- It has been confirmed around Earth—and is believed to be crucial near astrophysical black holes.
- The effect is strongest close to the black hole and in the equatorial plane.

"Spacetime is not a fixed stage—it moves with the dancers of the Universe."

Why It Matters

- Shows the deep connection between mass, rotation, and the geometry of spacetime.
- Essential for understanding black hole astronomy and relativistic jets.
- Provides a testbed for general relativity in the strong-field regime.
- Helps in the search for new physics beyond general relativity.

"A spinning black hole does more than bend spacetime—it drags it, twists it, and carries everything along for the ride."

10

The Ergosphere

The Region of Possible Energy Extraction

Outside the event horizon lies a remarkable region—the ergosphere—
where no object can remain at rest.
It is here that energy can be extracted from a rotating black hole.

What is the Ergosphere?

In a rotating (Kerr) black hole, frame dragging is so strong near the black hole that all particles are compelled to co-rotate. The ergosphere is the region outside the event horizon where this happens.

Ergosphere boundary (static limit):

$$r_{SL} = M + \sqrt{M^2 - a^2 \cos^2\theta}$$

Event horizon:

$$r_+ = M + \sqrt{M^2 - a^2}$$

For any θ,

$$r_+ \le r_{SL}$$

The ergosphere exists because $r_{SL} > r_+$, except at the poles.

Structure of a Rotating Black Hole

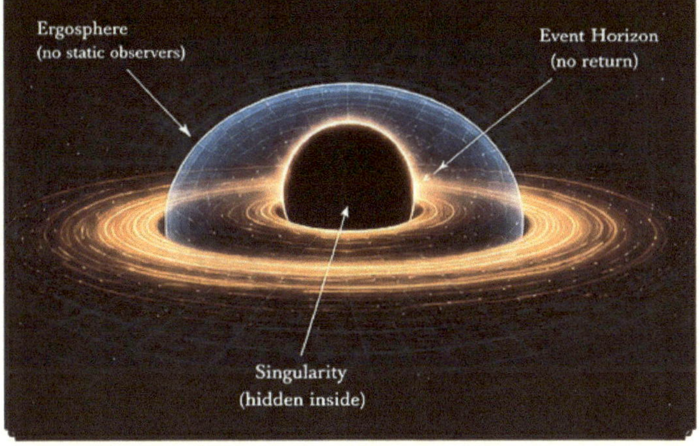

Key Facts

- Between r_+ and r_{SL} lies the ergosphere.
- No object can remain at rest here.
- All objects must move in the direction of rotation.
- Energy extraction is possible only within the ergosphere.
- At the poles ($\theta = 0, \pi$): $r_{SL} = r_+$, so the ergosphere touches the horizon.

For Extreme Rotation

($a \to M$)

At the equator ($\theta = \pi/2$):

$r_{SL} = 2M, \quad r_+ = M$

The ergosphere becomes very large.

No Rest in the Ergosphere

Inside the ergosphere, all timelike paths tilt. The "forward" direction in time forces objects to co-rotate.

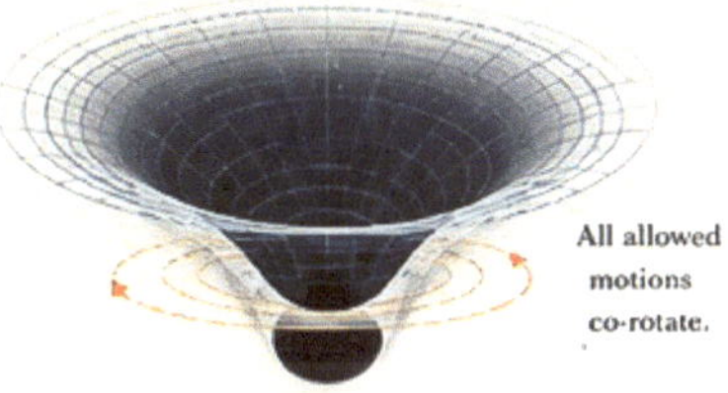

Inside the ergosphere, even "standing still" means moving.

Energy Extraction is Possible

A particle entering the ergosphere can split into two. One falls in with negative energy (relative to infinity), the other escapes with more energy than it had.

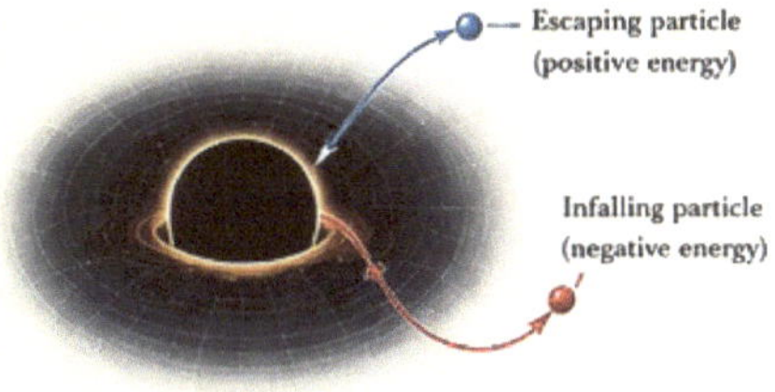

This is the essence of the Penrose process.

The Penrose Process (Overview)

Inside the ergosphere:

1. Particle 0 with energy E_0 enters.
2. It splits into particles 1 and 2.
3. Particle 2 falls in with $E_2 < 0$.
4. Particle 1 escapes with $E_1 > E_0$.
5. Black hole spin decreases.
6. Energy is extracted.

Energy conservation:

$$E_0 = E_1 + E_2 \quad \text{with} \quad E_2 < 0 \quad \Rightarrow \quad E_1 > E_0$$

Maximum efficiency (for $a \to M$):

$$\eta_{max} \approx 20.7\%$$

Shape of the Ergosphere

The ergosphere is oblate—flattened at the poles, bulging at the equator.

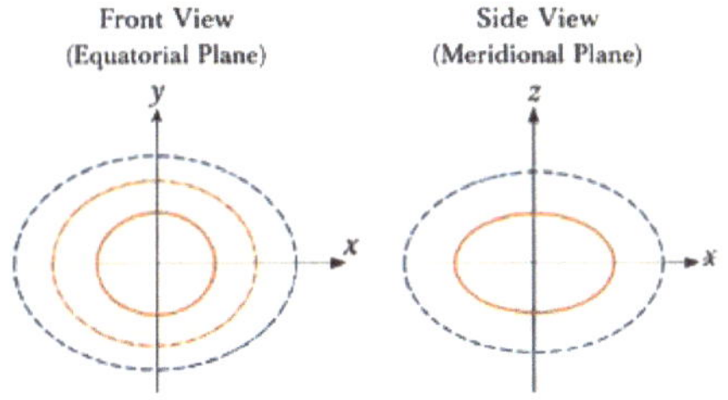

--- Static Limit ($r = r_{SL}$)

— Event Horizon ($r = r_+$)

For $a = 0$: no ergosphere.
For $a = M$: maximum bulge.

Size of the Ergosphere at the Equator ($\theta = \pi/2$)

As spin a increases, the ergosphere grows.

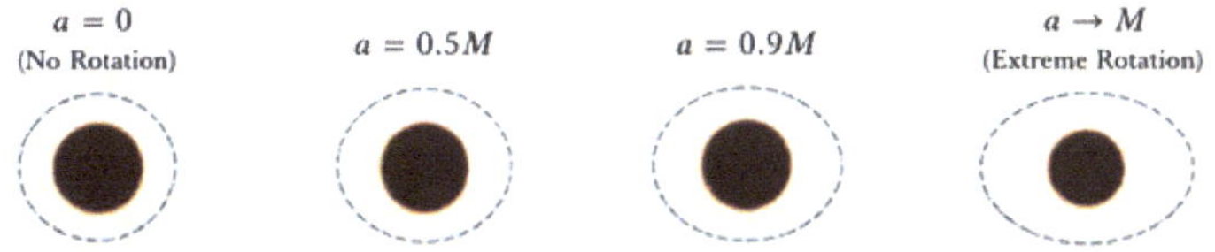

Spin a	0	0.5M	0.9M	0.99M	$\to M$
r_+/M	2.000	1.866	1.436	1.141	1.000
$r_{SL}(\theta = \pi/2)/M$	2.000	1.936	1.994	1.999	2.000
Ergosphere Width $r_{SL} - r_+$	0	0.070	0.558	0.8589	1.000

What Lives Inside the Ergosphere?

Accretion Flows

Plasma in the ergosphere inevitably co-rotates.

Angular momentum is dragged inward.

Magnetic Fields

Magnetic field lines are twisted by frame dragging.

Energy and angular momentum can be extracted (via BZ mechanism).

Waves

Electromagnetic and gravitational waves can extract energy.

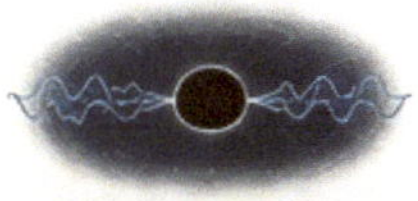

Basis of superradiance and wave amplification.

Negative Energy States

Only inside the ergosphere can negative-energy states exist (relative to infinity).

Key ingredient for energy extraction.

Boundaries Recap

— Event Horizon (r_+): No return.

---- Static Limit (r_{SL}): Boundary of ergosphere.

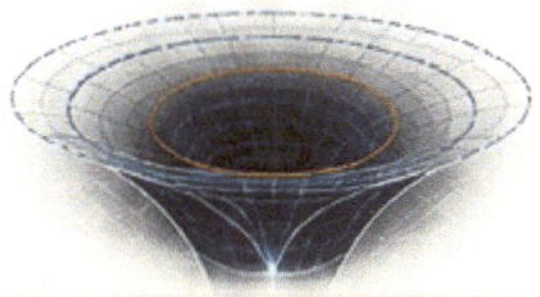

Key Takeaways

- The ergosphere is the region where spacetime itself is dragged.
- No object can remain at rest inside it.
- It is the only place where classical energy extraction is possible.
- Enables powerful astrophysical and theoretical phenomena.

> "The ergosphere is nature's engine room—
> a place where spacetime does the work,
> and energy can be set free."

Beyond the Ergosphere

From the ergosphere,
we move to the heart of black holes—
to the quantum world
and the final mystery.
→

The Ergosphere in Depth

A Reservoir of Rotational Energy

The ergosphere is not a physical surface—it is a region of spacetime where the very fabric is dragged around a rotating black hole.

1. What is the Ergosphere?

In a rotating (Kerr) black hole, the ergosphere is the region outside the event horizon where no object can remain at rest.

Boundaries (Boyer–Lindquist coords):

Event Horizon

$$r_+ = M + \sqrt{M^2 - a^2}$$

Ergosphere (static limit)

$$r_{SL} = M + \sqrt{M^2 - a^2 \cos^2\theta}$$

Between r_+ and r_{SL}, spacetime itself is compelled to rotate with the black hole. This is the ergosphere.

Note: $a = J/M, \quad 0 \le a \le M$

2. Structure of a Kerr Black Hole

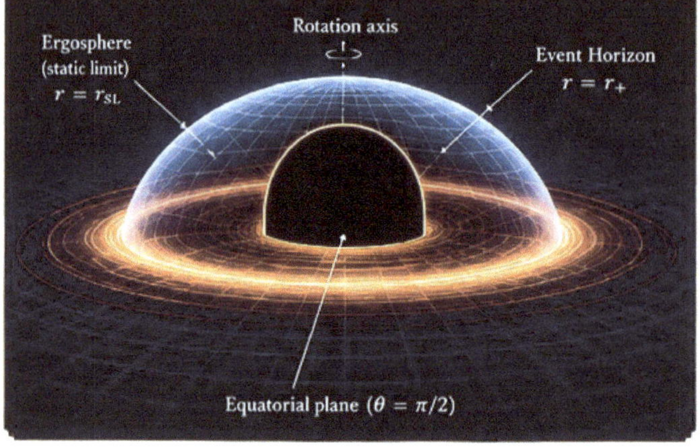

3. Key Facts

- Outside r_{SL}: objects can remain at rest (static observers exist).
- Inside ergosphere ($r_+ < r < r_{SL}$): no static observers—everything must co-rotate.
- Energy extraction is possible only in this region.
- The ergosphere is the source of the black hole's rotational energy.
- Maximum volume at equator ($\theta = \pi/2$).

Extreme Rotation ($a \to M$)
At the equator:
$r_{SL} = 2M, \quad r_+ = M$
The ergosphere becomes very large.

4. The "Impossible" Region

Inside the ergosphere, all future-directed timelike paths are forced to co-rotate.

- No object can hover.
- No object can stand still.
- Move forward in time—
- you must move in ϕ-direction.

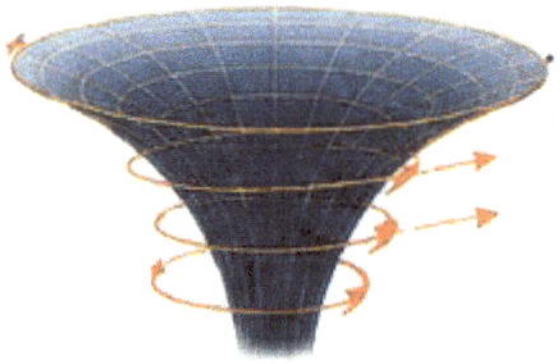

Light cones tip forward inside the ergosphere.

Toward the event horizon at the poles, but widest at the equator.

5. Negative Energy States

Inside the ergosphere, particles can have negative energy relative to infinity ($E < 0$).

Negative energy states cannot exist outside the ergosphere.

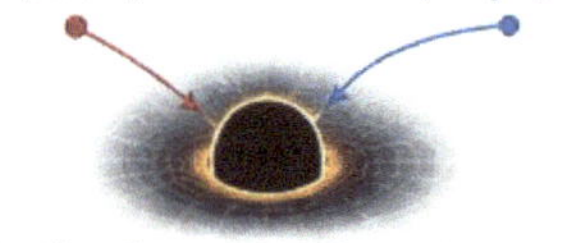

Key idea: if a particle with negative energy falls into the black hole, the escaping partner gains more energy than it had originally.

This is the basis of energy extraction!

6. Energy Extraction: The Penrose Process

1. Particle 0 with energy E_0 enters.
2. It splits into two particles 1 and 2.
3. Particle 1 falls in with $E_2 < 0$.
4. Particle 2 escapes with $E_1 > E_0$.
5. Black hole spin decreases.
6. Energy is extracted.

Energy conservation:
$E_0 = E_1 + E_2$ with $E_2 < 0 \quad \Rightarrow \quad E_1 > E_0$
(net energy extracted)

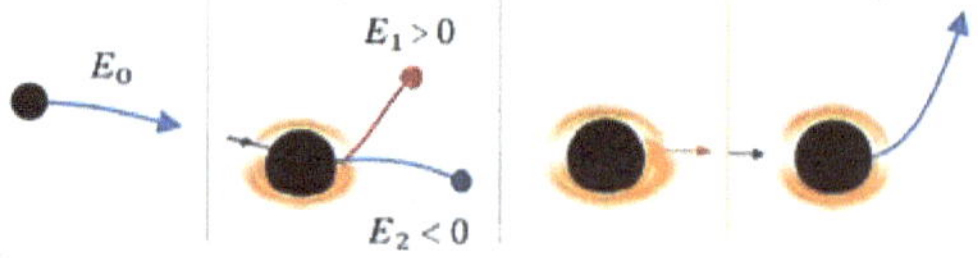

7. Geometry of the Ergosphere

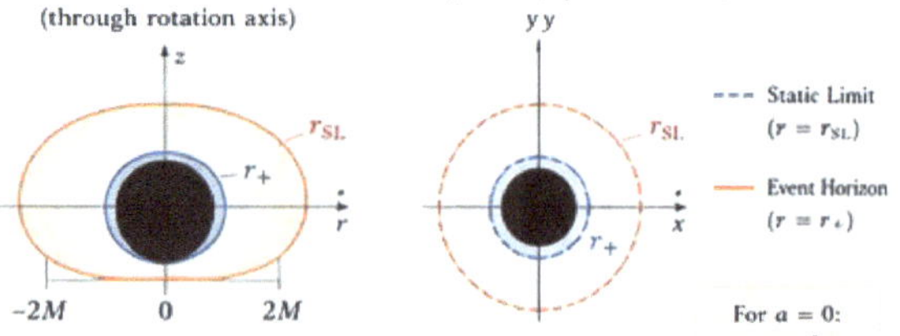

Oblate shape: flattened at poles, bulging at equator.

For $a = 0$: no ergosphere.
For $a \to M$: maximum bulge.

8. Size of the Ergosphere at the Equator ($\theta = \pi/2$)

As spin a increases, the ergosphere grows.

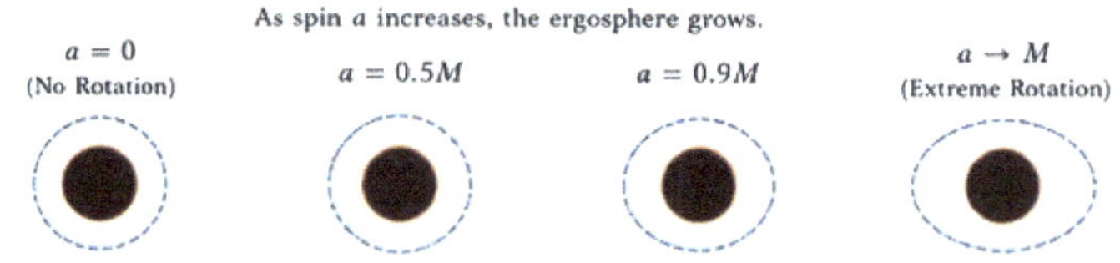

Spin a	0	$0.5M$	$0.9M$	$0.99M$	$\to M$
r_+/M	2.000	1.866	1.436	1.141	1.000
$r_{SL}\,(\theta = \pi/2)/M$	2.000	1.936	1.994	1.9999	2.000
Ergosphere Width $r_{SL} - r_+$	0	0.070	0.558	0.8589	1.000

The ergosphere volume increases rapidly with spin.

9. Observational Signatures

- Relativistic Jets (quasars, AGNs)
- X-ray Reflection (iron Kα line broadening)
- Shadow & Photon Ring (EHT images)
- Gravitational Waves (from mergers)

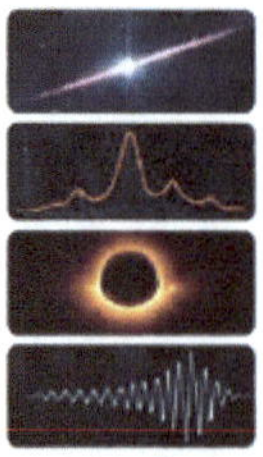

10. Key Takeaways

- The ergosphere exists only around rotating black holes.
- Spacetime itself is dragged—no rest allowed.
- Negative energy states inside allow energy extraction.
- This powers jets, flares, and some of the most energetic phenomena in the Universe.
- The faster the spin, the more energy the ergosphere can release.

11. Why it Matters

- Ergosphere is a natural battery built from rotation.
- It may explain the power of jets, gamma-ray bursts, and tidal disruption flares.
- Studying it tests general relativity in the strong-field regime.
- A key to unlocking the most extreme physics in the cosmos.

"The ergosphere is spacetime's engine room—where rotation becomes power, and free energy becomes possible."

12 QUANTUM VACUUM AND PAIR CREATION

HAWKING RADIATION: ENERGY FROM NOTHING?

Black holes are not completely black.
Quantum effects near the horizon allow antiparticle–particle pairs to form, and one particle escapes as radiation.

1. The Vacuum is Never Empty

According to quantum field theory, the vacuum is a seething sea of fluctuations. Virtual particle–antiparticle pairs constantly appear and annihilate in an extremely short time.

Pairs pop in and out of existence everywhere, all the time.

2. Near The Event Horizon

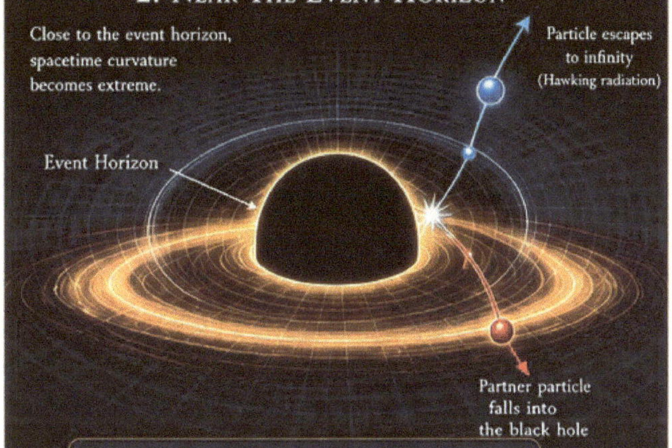

One particle escapes, the other is lost forever.
Energy is carried away from the black hole.

3. Hawking Radiation

The escaping particles form a thermal spectrum of radiation emitted by the black hole.

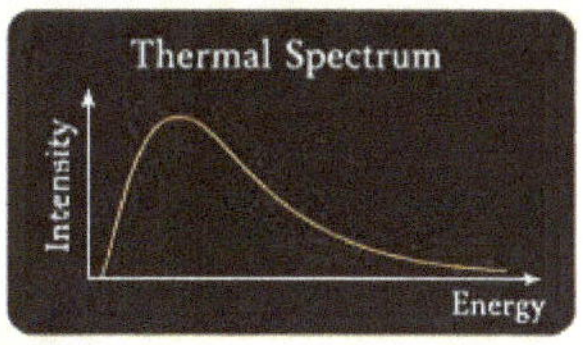

- Nearly perfect black bodies.
- Temperature inversely proportional to mass.
- Carries energy away, causing the black hole to evaporate slowly.
- Extremely weak for stellar black holes.

4. Pair Production Mechanism

1. A virtual pair appears near the horizon.
2. Quantum uncertainty allows separation.
3. One particle has negative energy (relative to infinity).
4. It falls into the black hole, reducing its mass.
5. The other has positive energy and escapes to infinity.

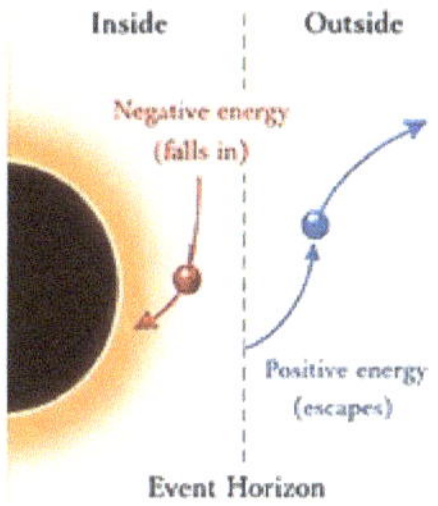

The black hole loses mass-energy E, but the total energy is conserved.

5. The Hawking Temperature

The temperature of a Schwarzschild black hole is

$$T_H = \frac{\hbar c^3}{8\pi G M k_B}$$

where

- $\hbar$ = reduced Planck constant
- c = speed of light
- G = gravitational constant
- M = mass of the black hole
- k_B = Boltzmann constant

Smaller black holes are hotter and evaporate faster.

6. Black Hole Evaporation

As radiation is emitted, the black hole shrinks and its temperature rises.

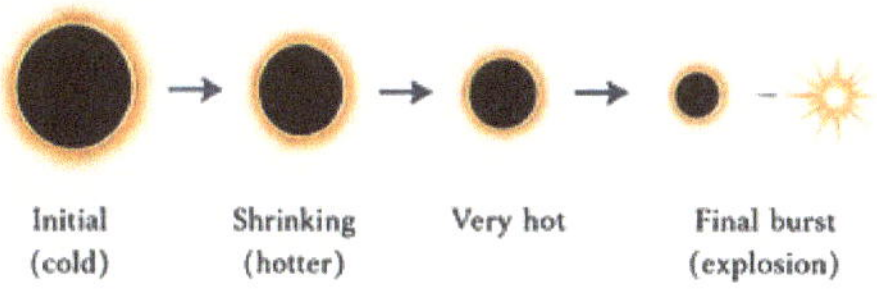

Eventually, the black hole may evaporate completely in a final burst of high-energy radiation (if no remnants remain).

7. Energy Conservation

The negative-energy particle reduces the black hole's mass. This ensures total energy is conserved.

Energy of escaping particle (positive) + Energy of fallen partner (negative) = Decrease in black hole mass

From the outside, only the escaping particle is seen, but the accounting balances.

8. Where Does the Energy Come From?

The energy comes from the gravitational field of the black hole itself.
Spacetime curvature provides the "fuel" for creating real particles.

Gravity does work on the quantum vacuum.
It is nature's way of paying the bill.

9. How Big is the Effect?

The effect is incredibly tiny for large black holes.

Black Hole Mass M	Temperature T_H	Evaporation Time
Stellar ($10\ M_\odot$)	$\sim 10^{-8}$ K	$\sim 10^{67}$ years
Supermassive ($10^9\ M_\odot$)	$\sim 10^{-13}$ K	$\sim 10^{90}$ years
Micro (10^{15} g)	$\sim 10^{12}$ K	$\sim 10^{-23}$ s (explodes)

Only very small (primordial) black holes could evaporate within the age of the Universe.

10. Why It Matters

- Hawking radiation unites quantum mechanics, thermodynamics, and gravity.
- It implies black holes have entropy:
  $$S = \frac{k_B A}{4\ell_P^2} \quad \text{(Bekenstein–Hawking entropy)}$$
- It leads to the black hole information paradox: Does information escape with the radiation?
- It hints that spacetime and quantum fields are deeply connected.

11. Key Takeaways

- The vacuum is full of virtual pairs.
- Near the horizon, pairs can split.
- One falls in with negative energy.
- The other escapes as Hawking radiation.
- Energy is extracted from the black hole.
- The process is tiny but profound.
- Black holes are not eternal—they can evaporate.

12. In a Nutshell

> *"Quantum effects near the horizon let nature borrow energy from gravity itself—turning nothing into something, and slowly erasing the black hole."*

"Hawking radiation shows us that even black holes have a quantum soul."

13 BIRTH OF HAWKING RADIATION

QUANTUM FLUCTUATIONS AT THE EDGE OF A BLACK HOLE

Hawking showed that quantum effects near the event horizon allow black holes to emit radiation—slowly losing mass and energy.

1. Quantum Fluctuations

Even in "empty" space, quantum theory predicts constant fluctuations.

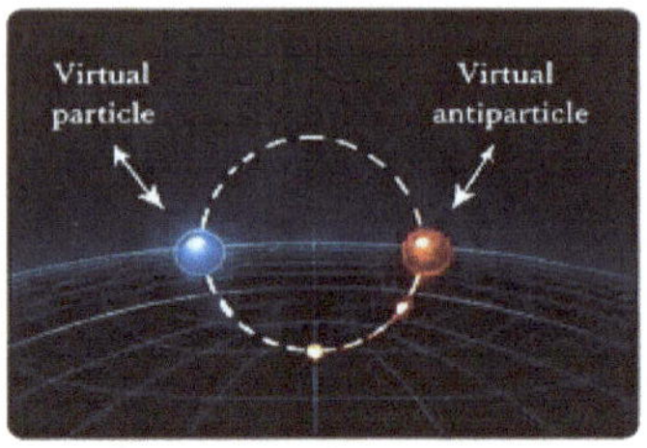

A pair of virtual particle–antiparticle appears for a short time and then annihilates.

These fluctuations happen everywhere in spacetime, all the time.

2. Near The Event Horizon

Near the event horizon, intense gravitational fields can tear a virtual pair apart.

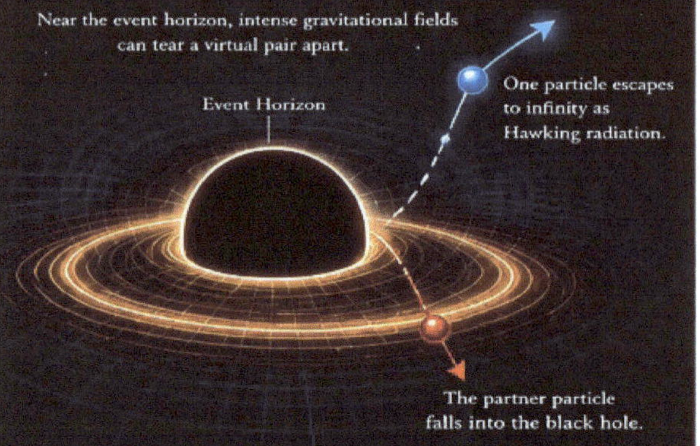

Energy is conserved: the black hole loses energy $E = mc^2$ when one particle escapes.

3. Hawking Radiation

The escaping particles form a thermal spectrum of radiation.

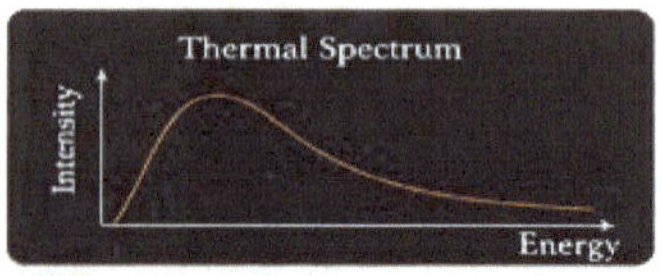

- Emitted particles are nearly blackbody radiation.
- Temperature is inversely proportional to mass:

$$T_H = \frac{\hbar c^3}{8\pi G M k_B}$$

- Lighter black holes are hotter.
- Radiation carries energy away.
- The black hole slowly evaporates.

4. The Hawking Temperature

For a non-rotating (Schwarzschild) black hole:

$$T_H = \frac{\hbar c^3}{8\pi G M k_B}$$

where

$\hbar$ = reduced Planck constant
c = speed of light
G = gravitational constant
M = mass of the black hole
k_B = Boltzmann constant

Example:
A solar-mass black hole has
$T_H \approx 6.2 \times 10^{-8}$ K
(very, very cold!)

5. Energy Loss and Evaporation

As Hawking radiation is emitted, the black hole loses mass.

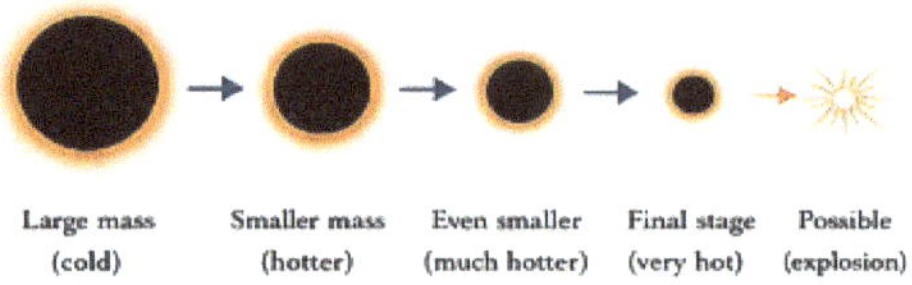

The evaporation accelerates as the black hole gets smaller. At the final stage, it may end in a high-energy burst (or leave a tiny remnant—still unknown).

6. Time Scales

Black Hole Mass M	Hawking Temperature T_H	Evaporation Time t
Stellar (10 $M_\odot$)	$\sim 10^{-8}$ K	$\sim 10^{67}$ years
Supermassive (10^9 $M_\odot$)	$\sim 10^{-13}$ K	$\sim 10^{103}$ years
Primordial (10^{12} kg)	$\sim 10^{11}$ K	$\sim 10^{-10}$ years
Tiny (10^{15} g)	$\sim 10^{13}$ K	$\sim$ 1 second

Only very small primordial black holes could evaporate within the current age of the Universe.

7. Pair Creation Picture

Inside: negative energy particle falls in.

Outside: positive energy particle escapes.

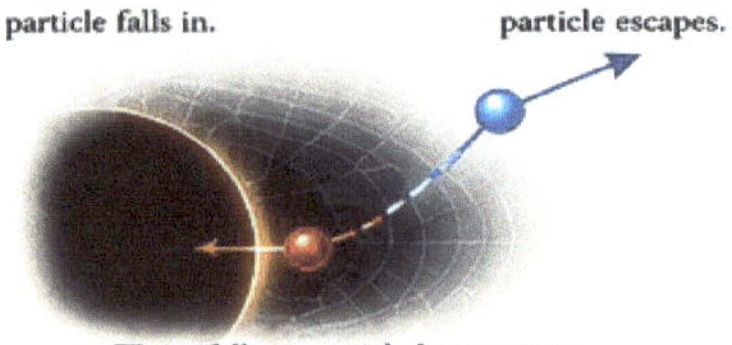

The infalling particle has negative energy relative to an outside observer.

This reduces the black hole's mass, allowing the other particle to escape.

8. From Virtual to Real

Virtual pairs are allowed by the uncertainty principle:

$$\Delta E \, \Delta t \geq \frac{\hbar}{2}$$

Near the horizon, Δt can be long enough that one member escapes, becoming real.

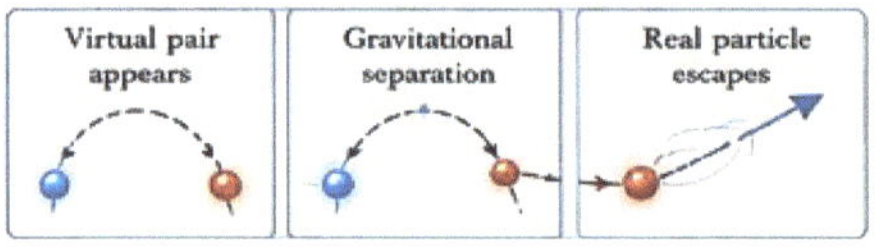

Hawking radiation is the quantum voice of the strong gravity at the horizon.

9. Key Implications

- Black holes are not completely black.
- They emit thermal radiation.
- They lose mass and energy over time.
- Information paradox arises: Does information escape with radiation?
- Connects quantum mechanics, thermodynamics, and gravity.

Hawking radiation was the first hint that quantum theory cannot be ignored in strong gravity.

10. Spectrum and Emission

The radiation is nearly perfect blackbody.

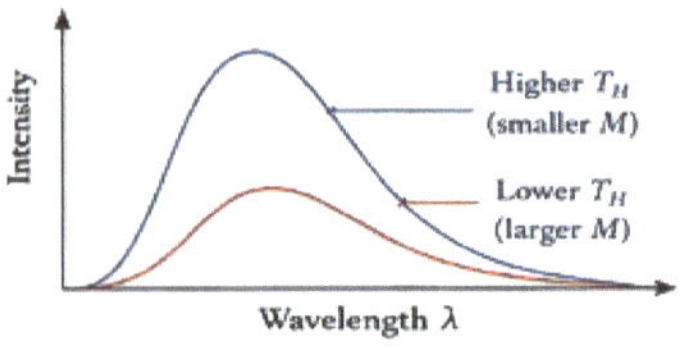

Peak shifts to shorter wavelengths as T_H increases.

11. Observational Prospects

- Astrophysical black holes are too cold to detect Hawking radiation.
- Primordial black holes could be hotter and might leave detectable signatures.
- Gamma-ray bursts or cosmic rays might carry traces.

Hawking radiation remains one of the greatest predictions waiting for experimental or astronomical confirmation.

12. Takeaway

> *"Quantum effects near the horizon let nature borrow energy from gravity itself—turning nothing into something, and slowly erasing the black hole."*

"Even black holes are not eternal—quantum mechanics gives them a voice, and that voice is Hawking radiation."

14

Black Hole Temperature

THE HOTTER THE SMALLER: A QUANTUM PARADOX OF GRAVITY

Hawking showed that black holes have a temperature inversely proportional to their mass. Smaller black holes are hotter and evaporate faster.

1. The Surprising Result

Classically, black holes absorb everything. Quantum theory changes the picture.

Black holes emit radiation like a blackbody!

This radiation has a temperature: the Hawking temperature.

2. Schwarzschild Black Hole

$$r_s = \frac{2GM}{c^2}$$

Schwarzschild radius

$$T_H = \frac{\hbar c^3}{8\pi G M k_B}$$

Hawking temperature

$\hbar$ = reduced Planck constant
c = speed of light
G = gravitational constant
M = mass of black hole
k_B = Boltzmann constant

Temperature is inversely proportional to mass: the smaller the black hole, the hotter it is.

3. Mass–Temperature Relation

A black hole's temperature drops as its mass increases.

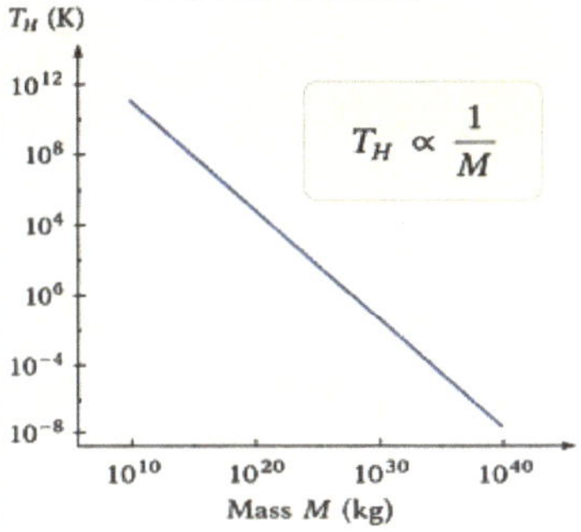

Stellar-mass black holes are extremely cold. Primordial black holes could be hot.

4. Temperature of Different Black Holes

Black Hole Type	Mass (kg)	Temperature T_H (K)
Stellar (10 $M_\odot$)	2×10^{31}	$\sim 10^{-8}$
Supermassive ($10^9\ M_\odot$)	2×10^{39}	$\sim 10^{-13}$
Primordial (10^{12} kg)	10^{12}	$\sim 10^{11}$
Primordial (10^{8} kg)	10^{8}	$\sim 10^{15}$
Primordial (10^{5} g)	10^{-8}	$\sim 10^{23}$

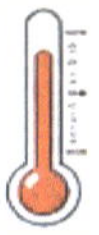

Primordial black holes with tiny mass could be extremely hot, even hotter than stars. They evaporate very quickly.

5. Blackbody Spectrum

Hawking radiation follows a blackbody spectrum with temperature T_H.

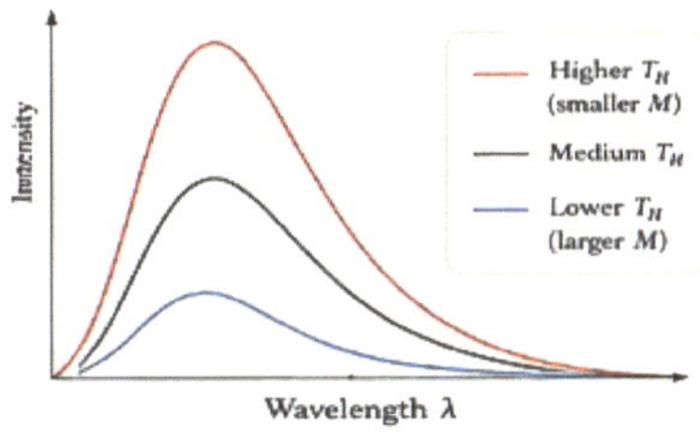

Hotter black holes emit more high-energy (short-wavelength) radiation.

6. Energy and Lifetime

Total power emitted:

$$P = \frac{\hbar c^6}{15360\pi G^2 M^2}$$

Lifetime (approx.):

$$\tau = \frac{5120\pi G^2 M^3}{\hbar c^4}$$

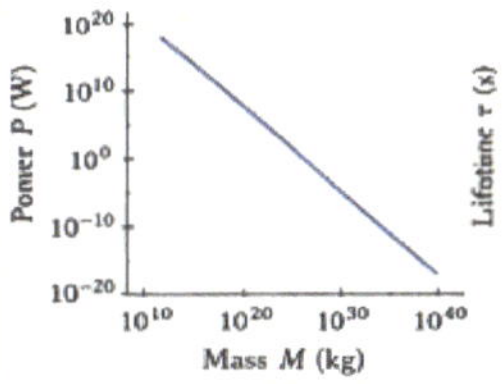

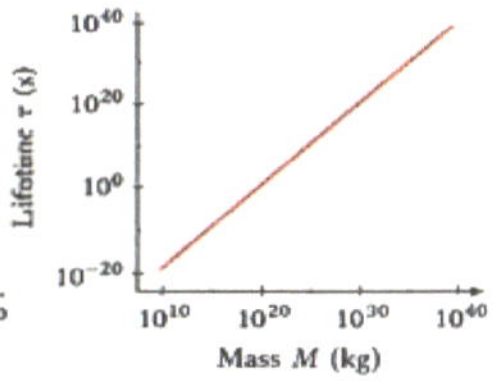

Smaller black holes radiate more power and have much shorter lifetimes.

7. Evaporation Process

As a black hole radiates, it loses mass. Its temperature rises—runaway evaporation.

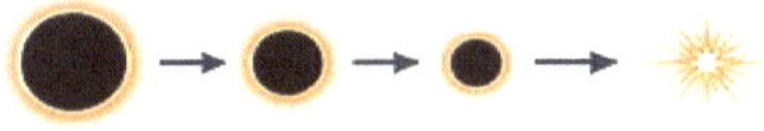

Stages:

1. Black hole emits Hawking radiation.
2. Mass decreases.
3. Temperature increases.
4. Evaporation speeds up.
5. Final explosion (Planck-scale end).

Only the smallest black holes evaporate within the age of the Universe.

8. Where Does the Energy Come From?

Quantum vacuum fluctuations near the horizon.

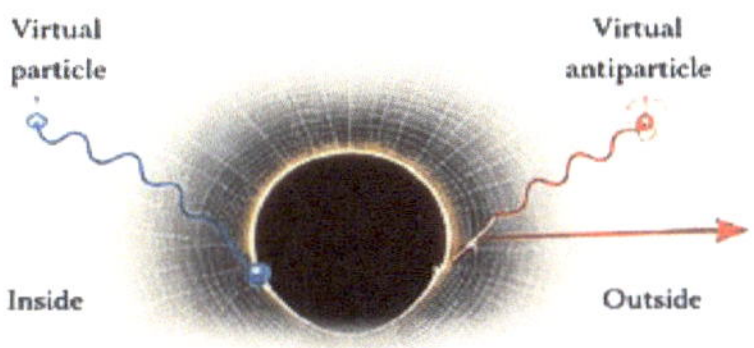

1. A virtual particle–antiparticle pair appears.
2. One falls into the black hole.
3. The other escapes and becomes real: Hawking radiation!

Energy is conserved: the black hole loses mass—energy.

9. Observational Prospects

Stellar and supermassive black holes: extremely cold → impossible to detect Hawking radiation.

Primordial black holes (if they exist): could be hot and may emit high-energy gamma rays.

Gamma-ray telescopes and cosmic ray detectors search for signatures.

No confirmed detection yet, but future observations may reveal it!

Hawking radiation bridges quantum theory, gravity, and the cosmos.

10. Key Equations Summary

- Schwarzschild radius: $r_s = \frac{2GM}{c^2}$
- Hawking temperature: $T_H = \frac{\hbar c^3}{8\pi G M k_B}$
- Power emitted: $P = \frac{\hbar c^6}{15360\pi G^2 M^2}$
- Lifetime: $\tau = \frac{5120\pi G^2 M^3}{\hbar c^4}$

11. Key Takeaways

- The vacuum is full of virtual pairs.
- Temperature is inversely proportional to mass.
- Smaller black holes are hotter.
- They lose mass and eventually evaporate.
- Quantum effects rule near the horizon.
- Hawking radiation unites quantum mechanics, thermodynamics, and gravity.

12. In a Nutshell

"Quantum fluctuations near the horizon allow black holes to emit radiation, lose energy, and eventually evaporate—revealing the deep unity of quantum theory and gravity."

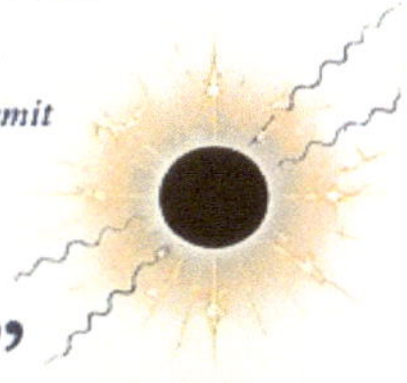

"Even the darkest objects glow—quantum whispers at the edge of spacetime."

15 ENTROPY AND THE AREA LAW

THE HIDDEN ORDER BEHIND BLACK HOLE THERMODYNAMICS

Black holes have entropy proportional to the area of their event horizon.
This profound result connects gravity, thermodynamics, and quantum theory.

1. Entropy: A Measure of Disorder

Entropy counts the number of microscopic states consistent with what we observe. More possbilities = higher entropy.

$$S = k_B \ln \Omega$$

S = entropy
k_B = Boltzmann constant
Ω = number of microstates

Order has few states. Disorder has many.

Inside a black hole, information disappears from view—but not from physics.

2. Black Hole Entropy: The Area Law

Bekenstein and Hawking discovered a remarkable fact:

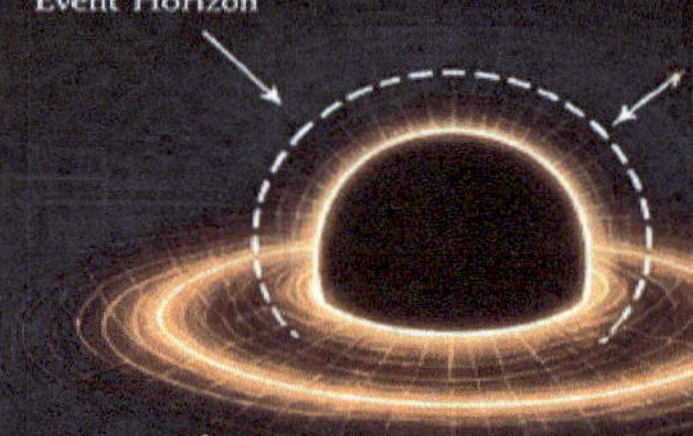

$$S_{BH} = \frac{k_B c^3 A}{4\hbar G}$$

Bekenstein–Hawking Entropy

A = area of the event horizon
$\hbar$ = reduced Planck constant
c = speed of light
G = gravitational constant

Entropy is proportional to area, not volume. Information resides on the surface.

3. Why Area, Not Volume?

Classically, entropy scales with volume. Black holes defy this intuition.

Normal Object (e.g., gas in a box)

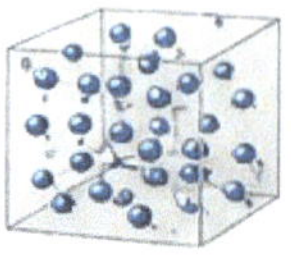

$S \propto$ Volume (more space, more states)

Black Hole

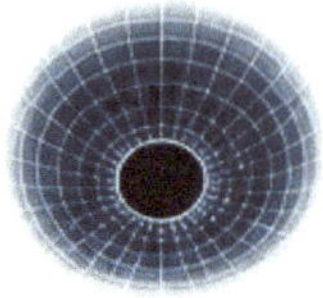

$S \propto$ Area (more surface, more states)

Gravity holographically encodes information on the boundary. Nature is more economical than we imagined.

4. Derivation Sketch

Hawking radiation → black hole temperature.
Thermodynamics → entropy.
First law of black hole thermodynamics:

$$dM = T_H dS + \Omega_H dJ + \Phi dQ$$

For a Schwarzschild black hole (non-rotating, uncharged):

$$dM = T_H dS$$

With $T_H = \frac{\hbar c^3}{8\pi G M k_B}$ and $M = \frac{rc^2}{2G}$; integration gives

$$S_{BH} = \frac{k_B c^3 A}{4\hbar G}$$

Thermodynamics and quantum theory reveal the area law.

5. Entropy and Black Hole Size

Entropy increases with horizon area. Larger black holes have much more entropy.

Black Hole Mass	Horizon Radius r_s	Area $A = 4\pi r_s^2$	Entropy S_{BH}
Stellar ($10\ M_\odot$)	~ 30 km	~ 1.1×10^{9} m²	~ $10^{77}\ k_B$
Supermassive ($10^{9}\ M_\odot$)	~ 3×10^{12} m	~ 1.1×10^{25} m²	~ $10^{93}\ k_B$
Primordial (10^{12} kg)	~ 1.5×10^{-15} m	~ 3×10^{-29} m²	~ $10^{24}\ k_B$

Stellar Black Hole

Supermassive Black Hole

Primordial Black Hole

Entropy spans an enormous range—from tiny primordial black holes to cosmic giants.

6. Horizon as an Information Screen

The horizon acts like a holographic screen. All information is encoded on its surface.

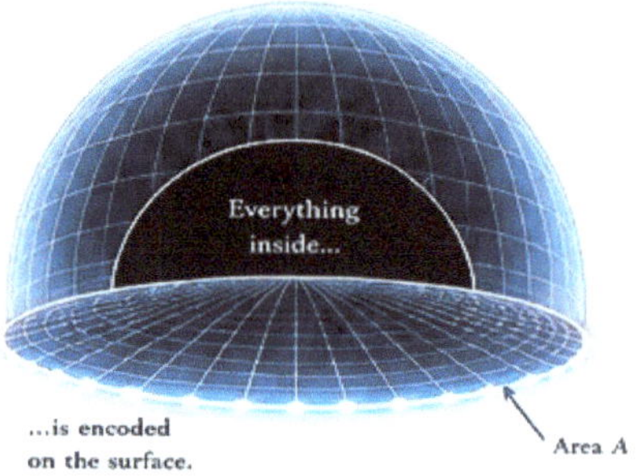

Holography: fewer dimensions can contain all the information.

7. The Generalized Second Law

The total entropy never decreases. Black hole + outside universe:

$$dS_{total} = dS_{outside} + dS_{BH} \geq 0$$

Even though the black hole loses mass, the total entropy always goes up.

8. Microstates and the Information Puzzle

What are the microstates behind black hole entropy? We still do not know for sure.

Possible Ideas:

- Quantum geometry
- String theory states
- Loop quantum gravity
- Horizon "atoms"
- Entanglement between degrees of freedom

Finding the microstates could solve the black hole information paradox.

9. Comparison with Other Systems

Other systems also have entropy, but none follow the area law.

System	Entropy Scaling	Example
Ideal gas	$S \propto$ Volume	Particles in a box
Radiation	$S \propto$ Volume	Photons in space
Black hole	$S \propto$ Area	Event horizon
Quantum spin chain	$S \propto$ Boundary	Entanglement entropy

Black holes are the ultimate example of holographic behavior.

10. Key Formulas

Bekenstein–Hawking entropy: $S_{BH} = \frac{k_B c^3 A}{4\hbar G}$

Hawking temperature (Schwarzschild): $T_H = \frac{\hbar c^3}{8\pi G M k_B}$

Schwarzschild radius: $r_s = \frac{2GM}{c^2}$

Horizon area: $A = 4\pi r_s^2$

These connect gravity, thermodynamics, and quantum theory.

11. Key Takeaways

- Black holes have entropy proportional to horizon area.
- This is a deep insight from quantum gravity.
- Smaller black holes are hotter and evaporate faster.
- The horizon acts as a holographic information screen.
- Total entropy of the universe never decreases.
- The microstates of black hole entropy remain a mystery.
- Understanding them may unlock quantum gravity.

12. Why it Matters

- Unifies gravity, thermodynamics, and quantum mechanics.
- Shapes our understanding of the cosmos and its ultimate laws.
- May solve the black hole information paradox.
- Hints that spacetime and information are deeply linked.

"The horizon hides the secret, but its surface speaks the language of the Universe."

Black Hole Evaporation

How Black Holes Fade Away

Hawking showed that black holes are not eternal: they radiate, lose mass, and can eventually disappear.

1. The Beginning of the End

A black hole forms when a massive star collapses. For a long time it is stable—but quantum effects change the story.

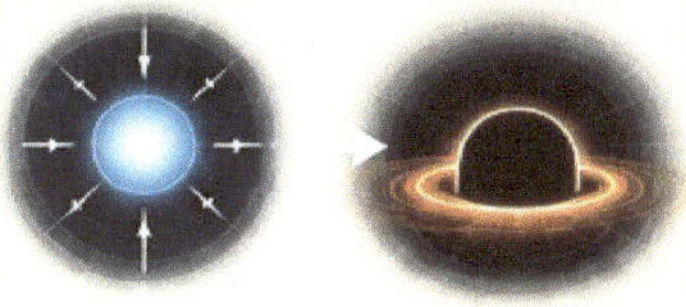

Star collapses — Black hole forms

Quantum fields near the horizon allow energy to leak out.

2. Pair Creation Near the Horizon

Virtual particle–antiparticle pairs pop into existence near the event horizon.

$$S_{BH} = \frac{k_B c^3 A}{4\hbar G}$$

Bekenstein–Hawking Entropy

- A = area of the event horizon
- $\hbar$ = reduced Planck constant
- c = speed of light
- G = gravitational constant
- k_B = Boltzmann constant

The escaping particle carries energy away. The black hole loses mass.

3. Thermal Hawking Radiation

The emitted radiation has a blackbody spectrum with temperature T_H.

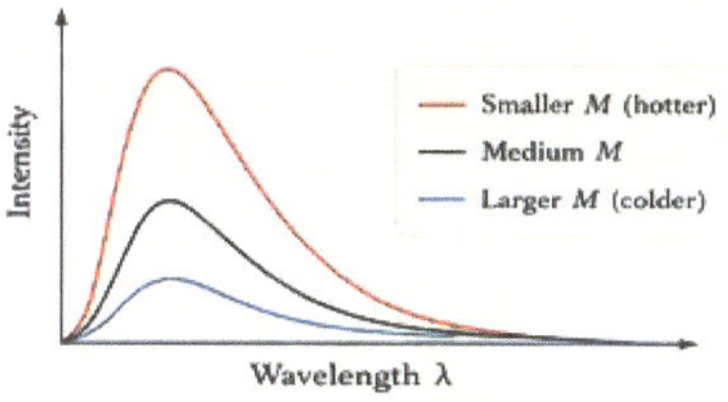

$$T_H = \frac{\hbar c^3}{8\pi G M k_B}$$

Smaller black holes are hotter and radiate more.

4. Mass Loss Over Time

As the black hole radiates, its mass decreases. The power emitted:

$$P = -\frac{dE}{dt} = \frac{\hbar c^6}{15360\pi G^2 M^2}$$

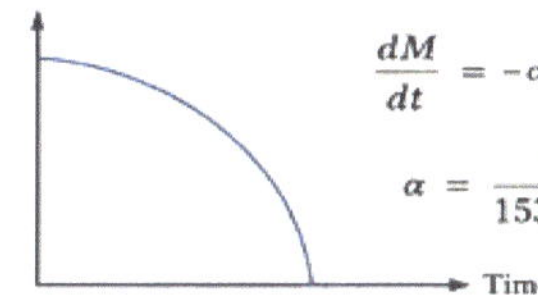

Mass evolution:

$$\frac{dM}{dt} = -\alpha\,\frac{1}{M^2}$$

$$\alpha = \frac{\hbar c^4}{15360\pi G^2}$$

The black hole evaporates completely in a finite time.

5. Lifetime of a Black Hole

The total evaporation time (from initial mass M_0):

$$\tau = \frac{5120\pi G^2}{\hbar c^4} M_0^3$$

Initial Mass M_0	Lifetime τ (approx.)	Example
Stellar (10 $M_\odot$)	$\sim 10^{67}$ years	Much longer than Universe
Supermassive (10^9 $M_\odot$)	$\sim 10^{94}$ years	Essentially eternal
Primordial (10^{12} kg)	$\sim 10^{10}$ years	May be exploding now
Tiny (10^{15} g)	$\sim 10^{-23}$ s	Instantaneous

Stellar Black Hole — Supermassive Black Hole — Primordial Black Hole

6. Evaporation Stages

The evaporation process has distinct phases.

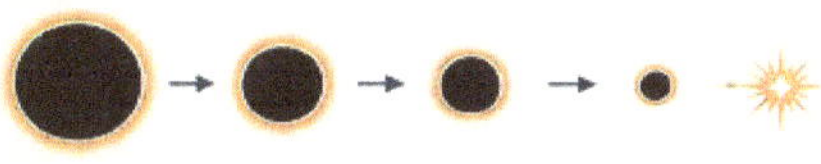

1. Early stage: black hole is large and cool.
2. Middle stage: mass shrinks, temperature rises.
3. Final stage: becomes extremely hot.
4. Final burst: ends in a high-energy explosion.

The last moments release a tremendous amount of energy.

7. Energy Conservation

Energy is conserved in the evaporation process.

Initial mass-energy of black hole $E_0 = Mc^2$ → Hawking radiation carried away → Final state (no black hole remains)

The total energy of radiation equals E_0.

Black holes do not violate energy conservation—they just return the energy slowly.

8. Entropy and Information

Black hole entropy is proportional to its horizon area $A = 4\pi r_s^2$.

$$S_{BH} = \frac{k_B c^3 A}{4\hbar G}$$

- As the black hole shrinks, its entropy decreases.
- If it evaporates completely, where does information go?
- This is the black hole information paradox.

The paradox drives research in quantum gravity.

9. Observational Implications

- Stellar and supermassive black holes are too long-lived to observe evaporating.
- Primordial black holes (if they exist) via their final bursts (gamma rays).
- Black hole evaporation could influence the early Universe.
- Future detectors may capture evaporation signatures.

We may one day witness a primordial black hole's final explosion.

10. Key Equations Summary

- Hawking temperature: $T_H = \frac{\hbar c^3}{8\pi G M k_B}$
- Power emitted: $P = \frac{\hbar c^6}{15360\pi G^2 M^2}$
- Mass loss rate: $\frac{dM}{dt} = -\frac{\alpha}{M^2}$, $\alpha = \frac{\hbar c^4}{15360\pi G^2}$
- Lifetime: $\tau = \frac{5120\pi G^2}{\hbar c^4} M_0^3$
- Black hole entropy: $S_{BH} = \frac{k_B c^3 A}{4\hbar G}$

These equations describe how black holes radiate, shrink, and disappear.

11. Key Takeaways

- Black holes are not eternal.
- Quantum effects near the horizon cause Hawking radiation.
- They lose mass and energy over time.
- Smaller black holes are hotter and evaporate faster.
- Evaporation completes in a finite time.
- The process connects gravity, quantum theory, thermodynamics, and information.

Black hole evaporation reveals the deep unity of the laws of nature.

12. The Big Picture

From birth to evaporation, a black hole's life follows the same story:

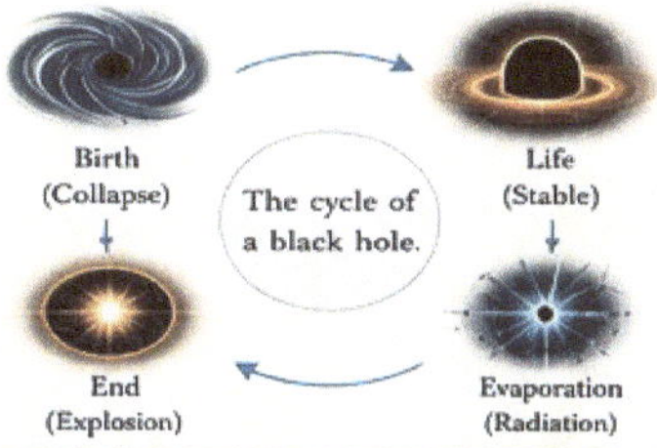

In the Universe, nothing truly disappears—it only changes form.

"Black holes teach us that even in the darkest places, quantum light can escape."

17 LIFETIME OF BLACK HOLES

HOW LONG DO BLACK HOLES LAST?

Hawking showed that black holes lose mass by radiation.
Their lifetime depends on their initial mass.

1. THE BIG SURPRISE

Classical theory predicted that black holes are forever. Quantum theory says otherwise.

Black holes emit Hawking radiation and slowly lose mass.

2. MASS EVOLUTION OVER TIME

As the black hole loses mass, its temperature increases.

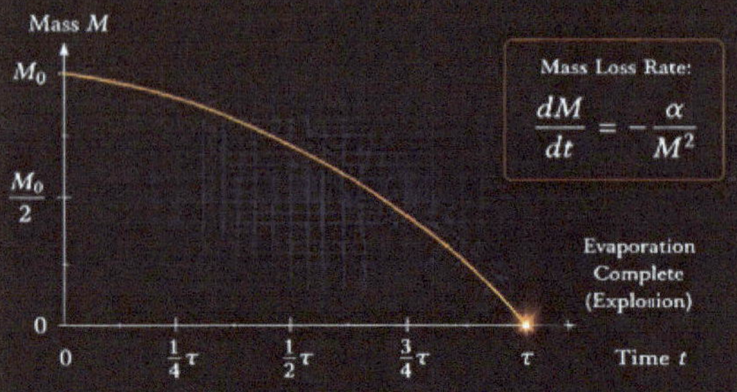

The mass decreases slowly at first, then very rapidly near the end.

3. TEMPERATURE VS MASS

Smaller black holes are hotter.
Temperature increases as mass decreases.

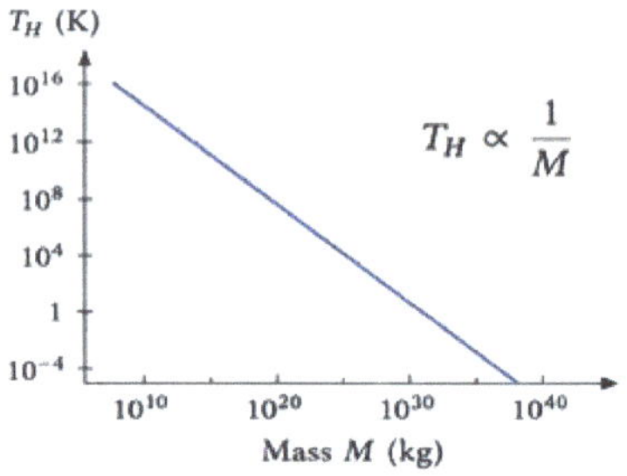

A tiny black hole can be hotter than a mountain of burning fire!

4. LIFETIME FORMULA

The total lifetime of a black hole with initial mass M_0 is:

$$\tau = \frac{5120\pi G^2}{\hbar c^4} M_0^3$$

where

- G = gravitational constant
- $\hbar$ = reduced Planck constant
- c = speed of light
- M_0 = initial mass

Lifetime is proportional to the cube of the mass! Huge black holes live much longer.

5. LIFETIME EXAMPLES

Lifetimes for different types of black holes:

Black Hole Type	Mass M_0 (kg)	Temperature T_H (K)	Lifetime τ (years)
Stellar ($10\ M_\odot$)	2×10^{31}	$\sim 10^{-8}$	$\sim 10^{67}$
Supermassive ($10^9\ M_\odot$)	2×10^{39}	$\sim 10^{-13}$	$\sim 10^{94}$
Primordial (10^{12} kg)	10^{12}	$\sim 10^{11}$	$\sim 10^{10}$
Primordial (10^{10} kg)	10^{10}	$\sim 10^{13}$	$\sim 10^{3}$
Primordial (10^{8} kg)	10^{8}	$\sim 10^{17}$	$\sim 10^{-7}$ s

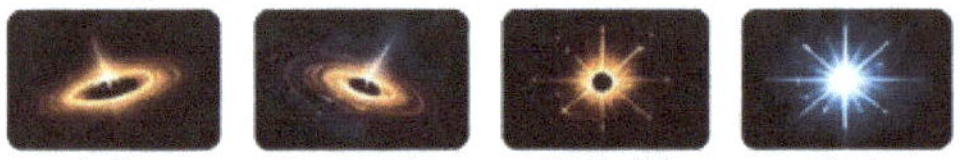

Stellar Black Hole | Supermassive Black Hole | Primordial (Small) | Final Explosion

Only the tiniest primordial black holes evaporate within the current age of the Universe.

6. EVAPORATION STAGES

The evaporation process has distinct phases.

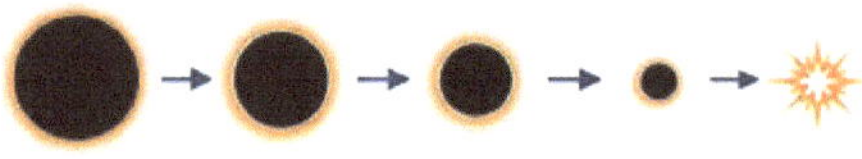

1. Early stage: black hole is large and cool.
2. Middle stage: mass shrinks, temperature rises.
3. Final stage: becomes extremely hot.
4. Final burst: ends in a high-energy explosion.

The last moments release a tremendous amount of energy.

7. POWER EMITTED

The power (energy per second) emitted:

$$P = -\frac{dE}{dt} = \frac{\hbar c^6}{15360\pi G^2 M^2}$$

Power increases as the black hole becomes smaller.

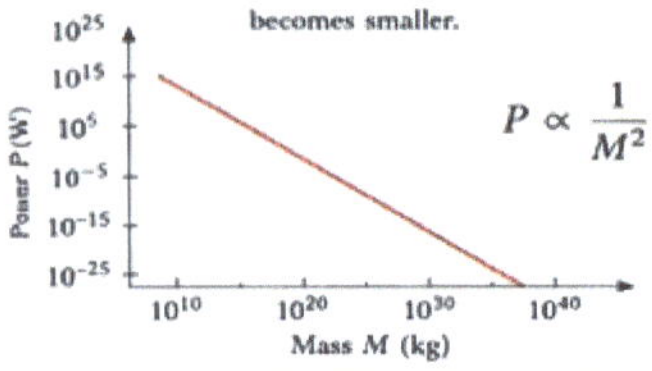

Tiny black holes are extremely powerful just before the end!

8. WHEN DO BLACK HOLES EVAPORATE?

Solving for lifetime:

$$\tau \approx 2.1 \times 10^{-17} \left(\frac{M_0}{\text{kg}}\right)^3 \text{ seconds}$$

Black holes with $M_0 \lesssim 5 \times 10^{11}$ kg evaporate within the current age of the Universe ($\sim 4.35 \times 10^{17}$ seconds).

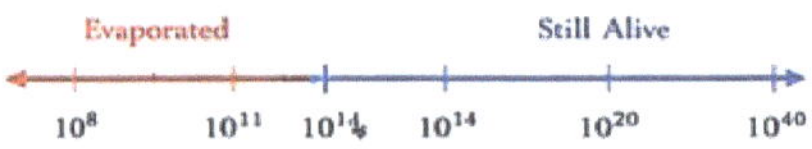

Smaller than $\sim 10^{11}$ kg: gone already.
Larger than that: still here.

9. OBSERVATIONAL IMPLICATIONS

- Gamma-ray bursts could be the final explosion of tiny primordial black holes.
- Searches for these signals help test early Universe models.
- Black hole evaporation connects the very large (cosmos) and very small (quantum).
- No confirmed detection yet— but the possibility remains exciting!

Black hole evaporation may be a window into the earliest moments of the Universe.

10. KEY EQUATIONS SUMMARY

- Hawking Temperature: $T_H = \frac{\hbar c^3}{8\pi G M k_B}$
- Power Emitted: $P = \frac{\hbar c^6}{15360\pi G^2 M^2}$
- Mass Loss Rate: $\frac{dM}{dt} = -\frac{\alpha}{M^2} \approx -\frac{\hbar c^4}{15360\pi G^2 M^2}$
- Lifetime: $\tau = \frac{5120\pi G^2}{\hbar c^4} M_0^3$
- Black Hole Entropy: $S_{BH} = \frac{k_B c^3 A}{4\hbar G}$

These equations describe how black holes radiate, shrink, and disappear.

11. KEY TAKEAWAYS

- Black holes are not eternal.
- They emit Hawking radiation.
- Smaller black holes are hotter and evaporate faster.
- They lose mass and energy over time.
- Only tiny primordial black holes evaporate within today's Universe.
- Evaporation ends in a high-energy explosion.
- This links quantum theory with gravity.

Black holes live, radiate, and die— quantum mechanics gives them a life cycle.

12. IN A NUTSHELL

> *"Black holes are like cosmic candles: they shine with quantum light, grow dim as they lose mass, and finally burn out in a brilliant flash."*

"From nothing but quantum whispers, even black holes can disappear."

18

The Penrose Process

Extracting Energy from a Rotating Black Hole

Before Hawking radiation was discovered, Roger Penrose showed that rotating black holes can lose energy through purely classical physics.

1. Rotation Creates an Ergosphere

A rotating (Kerr) black hole drags spacetime around with it. Outside the event horizon lies the ergosphere.

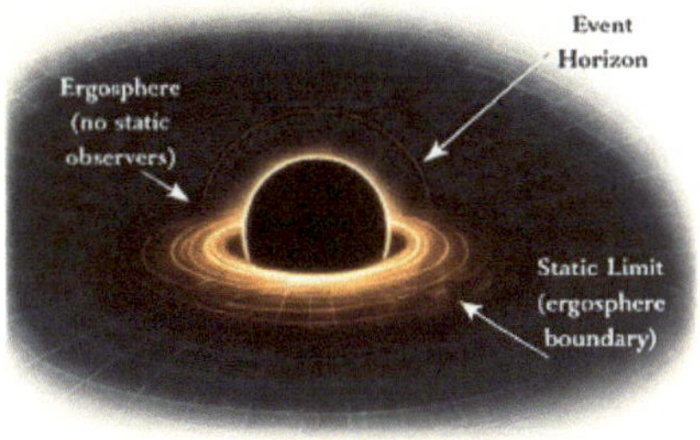

Inside the ergosphere, all objects are forced to co-rotate with the black hole.

2. Splitting a Particle

A particle enters the ergosphere and splits into two fragments.

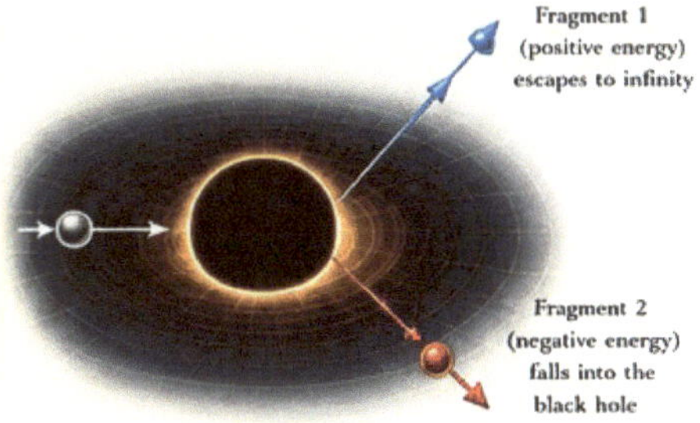

Due to frame dragging, one fragment can have negative energy relative to infinity.

3. Energy Extraction

The escaping fragment carries away more energy than the original particle had.

$$E_{out} > E_{in}$$

The black hole's rotational energy decreases.

Energy is extracted, and the black hole spins down.

Conditions for the Process

- The black hole must be rotating ($a \neq 0$).
- The splitting occurs inside the ergosphere.
- One fragment must have negative energy.
- The negative-energy fragment falls in.
- The positive-energy fragment escapes.

4. How Negative Energy is Possible

Inside the ergosphere, frame dragging is so strong that all particles must move in the direction of rotation.

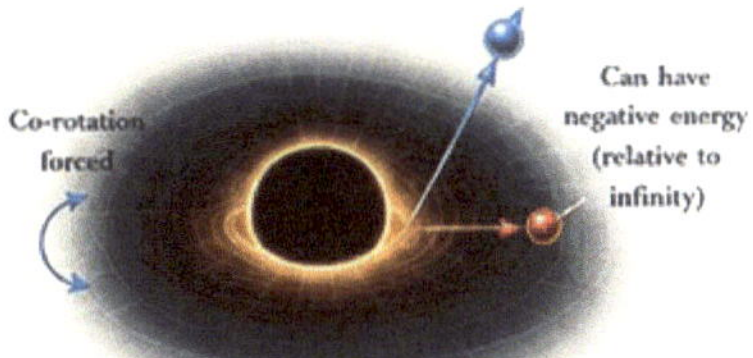

A particle moving backward (against rotation) can still be dragged forward—giving it negative energy.

5. Maximum Extractable Energy

The extractable energy depends on the black hole's spin parameter a ($0 \leq a < M$).

Spin a/M	Max Extractable Energy (% of Mc^2)	Final Spin (a_f/M)
0 (non-rotating)	0 %	0
0.5	3.4 %	~ 0
0.7	8.6 %	~ 0
0.9	15.3 %	~ 0
0.99	23.4 %	~ 0
1 (extremal)	29.3 %	0

In principle, up to 29.3% of the black hole's mass–energy can be extracted.

6. Visual Summary

The Penrose process is purely classical—no quantum effects needed.

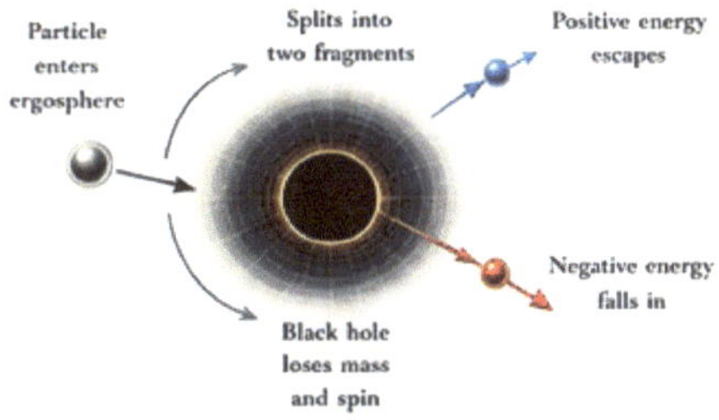

A beautiful interplay of gravity, rotation, and relativity.

7. Mathematical Picture

The conserved energy of a particle (at infinity) is

$$E = -p_t.$$

Inside the ergosphere: E can be negative!

Penrose condition:

$$0 < E_{out} < E_{in}, \quad E_{in} = E_{out} + E_{infall}$$

$$\text{with} \quad E_{infall} < 0$$

The black hole's mass decreases by $|E_{infall}|$.

8. Astrophysical Relevance

The Penrose process may operate in nature:

Near rapidly rotating black holes with strong accretion disks.

Could help power relativistic jets in quasars and active galaxies.

Works with or without magnetic fields (unlike Blandford–Znajek mechanism).

A key step in understanding how black holes power the most energetic phenomena in the Uniories.

9. Penrose Process vs. BZ Mechanism

Feature	Penrose Process (Classical)	Blandford–Znajek (BZ) (Electromagnetic, MHD)
Nature	Particle (classical)	Electromagnetic, MHD)
Requires magnetic field?	No	Yes
Energy source	Black hole rotation	Black hole + magnetic field
Where it occurs	Ergosphere	Ergosphere + field lines
Mechanism	Particle splitting	Magnetic extraction
Discovered by	Roger Penrose (1969)	Blandford & Znajek (1977)

Both extract rotational energy—by different routes.

10. Key Equations

- Ergosphere condition (Kerr metric):

$$\Delta - 2Mr < 0$$

- Particle energy at infinity:

$$E = -p_t$$

- Penrose condition:

$$E_{infall} < 0, \quad E_{out} > E_{in}$$

- Max extractable energy (theoretical):

$$E_{max} = \left(1 - \frac{1}{\sqrt{2}}\right) Mc^2 \approx 0.293\, Mc^2$$

- The deeper the spin, the more energy can be extracted.

11. Key Takeaways

- Rotating black holes have an ergosphere.
- Inside it, negative energy states are possible.
- Particle splitting can extract energy.
- The black hole loses mass and angular momentum.
- Up to 29.3% of the mass–energy can be extracted.
- The Penrose process is classical and fundamental.
- It inspires modern astrophysics and jet theories.

A half-century-old idea that still powers today's frontiers.

12. In One Sentence

> *"The Perrose process shows that rotation and relativity combine to let a black hole literally give up part of its energy—by letting one piece fall in with a debt, so another can escape with a profit."*

In the dance of gravity and rotation, even black hole can be made to give."

19

Negative Energy Orbits

The Fuel Behind Energy Extraction

Inside the ergosphere, space itself is dragged so strongly by the black hole's rotation that some orbits can have negative energy—allowing the Penrose process to work.

1. The Ergosphere Revisited

Outside the ergosphere, all particle orbits have positive energy. Inside, some can have negative energy relative to infinity.

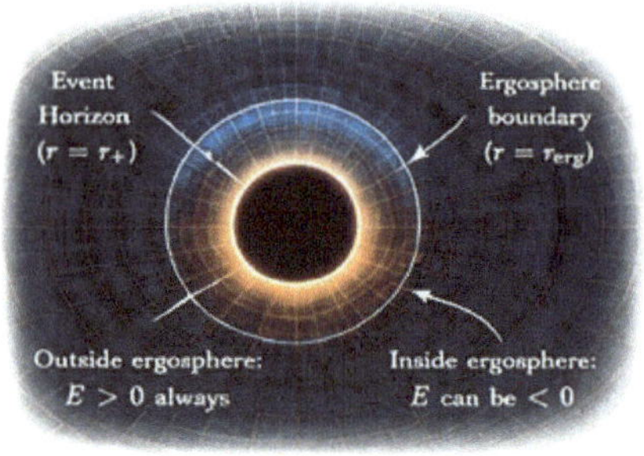

The ergosphere is the key: it allows negative energy states to exist.

2. Energy of Particle Orbits

The energy (E) of a test particle in Kerr spacetime (as measured at infinity) is

$$E = -p_t$$

For equatorial motion ($\theta = \pi/2$):

$$E = \frac{(r^2 + a^2)\,\dot{t} - a\,\dot{\phi}\,(2Mr)}{\Sigma}$$

where $\Sigma = r^2 + a^2\cos^2\theta$ and dots are derivatives with respect to proper time τ.

Inside the ergosphere, the frame dragging (term with $\dot{\phi}$) can make E negative.

3. Where can E be Negative?

For prograde orbits inside the ergosphere and near the event horizon, the energy can be negative.

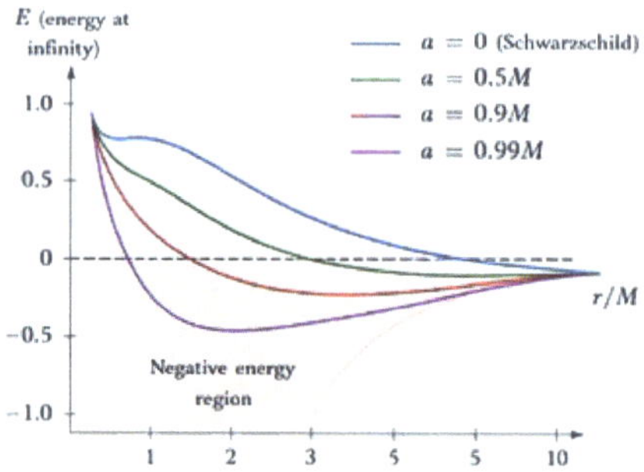

Higher spin (a closer to M) makes the negative energy region larger and deeper.

4. Example: Prograde Orbits

For a rapidly rotating black hole ($a = 0.99M$), particles on prograde orbits can reach $E < 0$ inside the ergosphere.

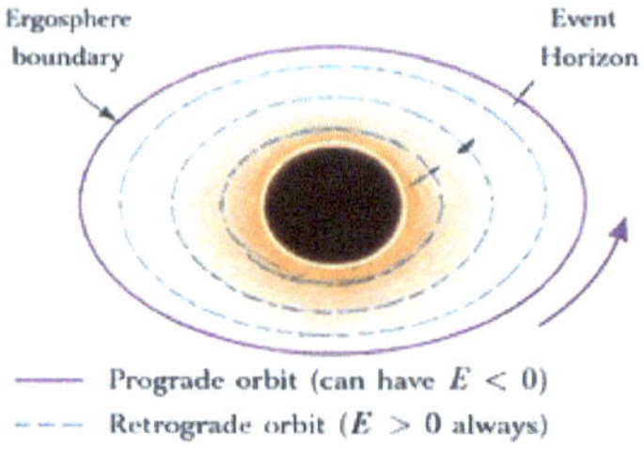

Only prograde orbits close to the horizon can access negative energy.

5. Negative Energy is Not "Strange"

The negative energy is not "negative mass"—it is energy defined relative to infinity.

Outside ergosphere

- Killing vector ∂_t is timelike
- Energy is positive definite: $E > 0$

Inside ergosphere

- ∂_t becomes spacelike
- Energy can be negative: $E < 0$
- Allowed by general relativity!

General relativity allows negative energy states in curved spacetime (with conditions).

6. How Negative Energy Helps

If one fragment falls in with $E < 0$, the other can escape with $E_{out} > E_{in}$—a net energy gain!

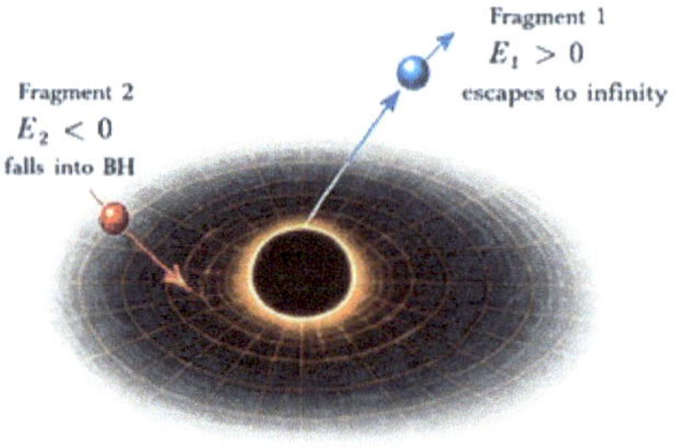

The black hole's rotational energy decreases, but energy is extracted to infinity.

7. Bounds on Negative Energy

There is a limit to how negative the energy can be.

$$E_{min} = \frac{\sqrt{3}\,\Omega_H}{1 + \sqrt{1 - a^2/M^2}}$$

where $J = aM$ is the angular momentum and $\Omega_H = \frac{a}{2Mr_+}$ is the horizon's angular velocity.

For $a = 0.99M$: $E_{min} \approx -0.19mc^2$

Negative energy exists, but within strict limits determined by the black hole's spin.

8. Intuition: "Borrowing" Rotation

A particle can give back more rotational energy than it brought in.

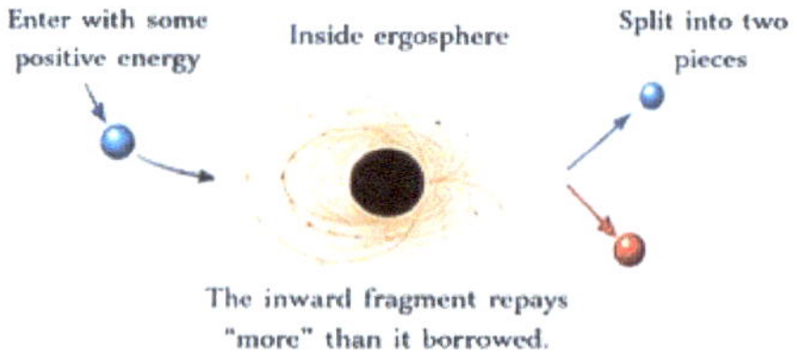

This black hole's rotation does the extra work.

9. Connection to Angular Momentum

Negative energy orbits always carry angular momentum in the same direction as the black hole's spin.

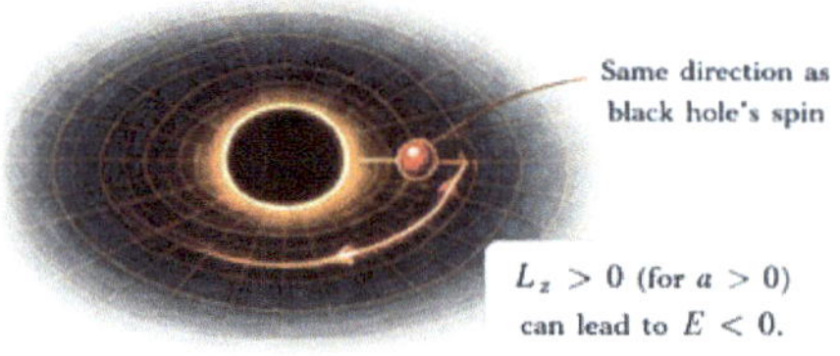

Counter-rotating orbits cannot achieve $E < 0$.

10. Key Equations Summary

- Energy at infinity: $E = -p_t$
- Equatorial energy: $E = \frac{(r^2 + a^2)\,\dot{t} - a\,\dot{\phi}\,(2Mr)}{\Sigma}$
- Horizon angular velocity: $\Omega_H = \frac{a}{2Mr_+}$
- Minimum energy (prograde): $E_{min} \approx -0.19mc^2$ (for $a = 0.99M$)
- Max extractable energy (theoretical): $E_{max} = \left(1 - \frac{1}{\sqrt{2}}\right) Mc^2 \approx 0.293\,Mc^2$

These relations govern negative energy orbits inside the ergosphere.

11. Key Takeaways

- Rotating black holes have an ergosphere.
- Inside it, negative energy states are possible.
- Particle splitting can extract energy.
- The black hole loses mass and angular momentum.
- Up to 29.3% of the mass–energy can be extracted.
- The Penrose process is classical and fundamental.
- Black holes can be tapped for rotational energy through classical physics!

Negative energy orbits are a doorway to harnessing a black hole's spin.

12. In One Sentence

> "*Inside the ergosphere, rotating space allows orbits with negative energy—so the black hole can give up part of its spin and power the cosmos.*"

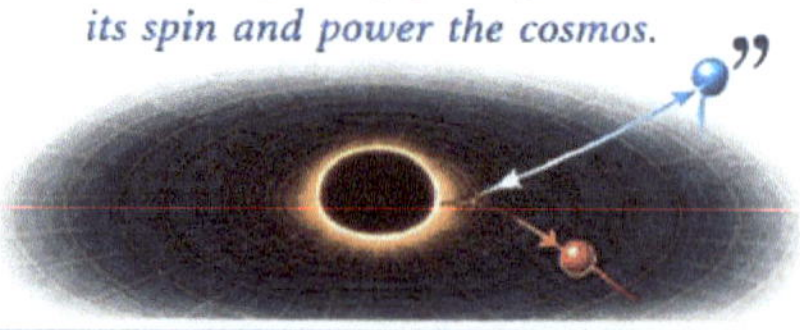

"Negative energy is not a paradox—it is nature's way of letting black holes share their hidden power."

20 Maximum Energy Extraction Efficiency

How Much Energy Can We Get?

The Penrose process has a limit. How much rotational energy can a black hole give up? The answer depends only on its spin.

1. Rotational Energy Reservoir

A rotating (Kerr) black hole has rotational energy stored in its spin.

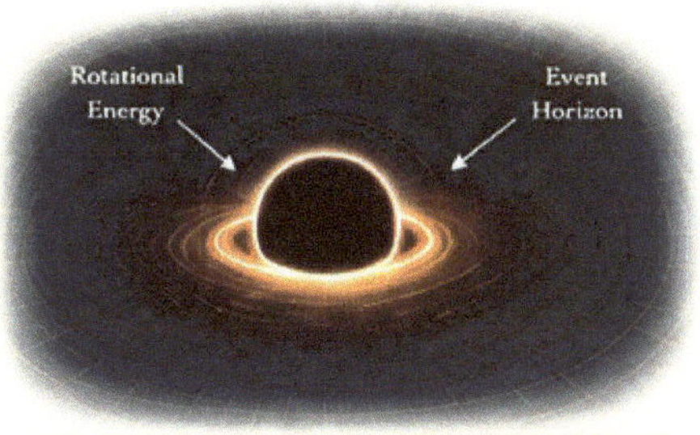

More spin → more rotational energy available to be extracted.

2. Black Hole Mass–Energy Split

The total mass–energy M of a Kerr black hole can be written as:

$$M^2 = M_{\text{irr}}^2 + M_{\text{rot}}^2$$

where

M_{irr} = irreducible mass (cannot be extracted)

M_{rot} = rotational mass–energy (extractable)

$$M_{\text{irr}} = \frac{1}{2} M \left(1 + \sqrt{1 - a^2}\right)$$

$$M_{\text{rot}} = M - M_{\text{irr}}$$

$$a = \frac{J}{M^2} \quad (0 \le a < 1)$$

J = angular momentum

Only M_{rot} can be extracted by any process that respects ordinary physics.

3. Maximum Extractable Energy

The maximum possible efficiency $\eta_{\max}$ is the fraction of M that can be extracted:

$$\eta_{\max} = \frac{M_{\text{rot}}}{M} = 1 - \frac{M_{\text{irr}}}{M}$$

Therefore,

$$\eta_{\max} = 1 - \frac{1 + \sqrt{1 - a^2}}{2}$$

Simplified:

$$\eta_{\max} = \frac{1 - \sqrt{1 - a^2}}{2}$$

This is the absolute upper limit for energy extraction from a Kerr black hole.

4. Efficiency vs Spin

Higher spin means higher maximum efficiency.

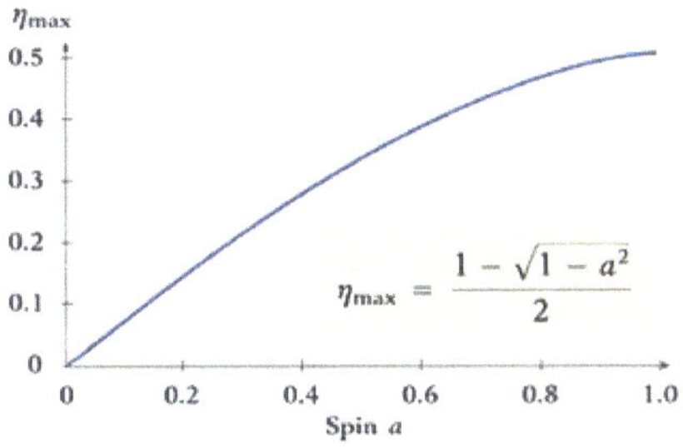

A maximally rotating black hole ($a \to 1$) can give up nearly 50% of its mass–energy!

5. Numerical Values

Some important cases:

Spin a	$\sqrt{1 - a^2}$	$\eta_{\max}$ (fraction)	$\eta_{\max}$ (%)
0 (Schwarzschild)	1	0	0 %
0.5	0.8660	0.06699	6.70 %
0.7	0.7141	0.1429	14.29 %
0.9	0.4359	0.2821	28.21 %
0.95	0.3122	0.3439	34.39 %
0.99	0.1411	0.4295	42.95 %
0.998	0.0632	0.4684	46.84 %
1 (extremal)	0	0.5000	50.00 %

As $a \to 1$, the efficiency approaches 50%, but never exceeds it.

6. Where Does the Limit Come From?

The irreducible mass M_{irr} is tied to the horizon area:

$$A = 16\pi M_{\text{irr}}^2$$

Hawking's area theorem:

$$\frac{dA}{dt} \ge 0 \quad \text{(classically)}$$

The horizon area cannot decrease in any physical process.

Therefore, M_{irr} cannot decrease as long as the area holds.

No process can extract more than M_{rot} without violating the laws of physics.

7. Comparison of Extraction Mechanisms

All classical mechanisms share the same limit.

Mechanism	How it Works	Max Efficiency $\eta_{\max}$
Penrose Process	Particle splitting inside ergosphere	$\eta_{\max}$
Superradiance	Wave amplification (modes)	$\eta_{\max}$
Blandford–Znajek	EM fields + magnetic field lines	$\eta_{\max}$

Different methods, same ultimate limit—set by the black hole's spin.

8. Extremal Case ($a \to 1$)

As a approaches 1:

- $M_{\text{irr}} \to \frac{1}{2} M$
- $M_{\text{rot}} \to \frac{1}{2} M$
- $\eta_{\max} \to 50\,\%$

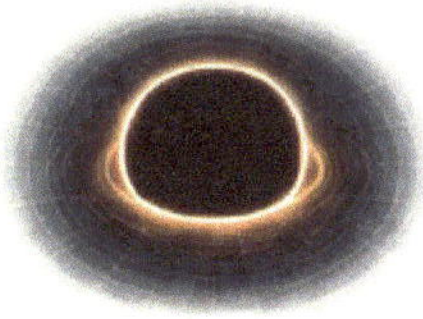

A maximally rotating black hole can give up half its mass–energy.

But $a = 1$ is not attainable in practice—radiation, matter, and instabilities prevent it.

9. Connection to Angular Momentum

Negative energy orbits always carry angular momentum in the same direction as the black hole's spin.

$L_z > 0$ (for $a > 0$) can lead to $E < 0$.

Counter-rotating orbits cannot achieve $E < 0$.

10. Key Formulas

- Mass decomposition:

$$M^2 = M_{\text{irr}}^2 + M_{\text{rot}}^2$$

- Irreducible mass:

$$M_{\text{irr}} = \frac{1}{2} M \left(1 + \sqrt{1 - a^2}\right)$$

- Maximum extractable energy:

$$E_{\max} = M_{\text{rot}} = M - M_{\text{irr}}$$

- Maximum efficiency:

$$\eta_{\max} = \frac{1 - \sqrt{1 - a^2}}{2}$$

These relations govern how much energy can be extracted.

11. Key Takeaways

- Rotating black holes have an energy reservoir.
- Inside it, negative energy states are possible.
- Particle splitting can extract energy.
- The black hole loses mass and angular momentum.
- Up to 29.3% of the mass–energy can be extracted.
- The Penrose process is classical and fundamental.
- All classical extraction mechanisms share the same limit.
- Understanding this limit is crucial for astrophysics, black hole energetics, and high-energy phenomena.

Negative energy orbits are a doorway to harnessing a black hole's spin.

12. In one Sentence

"A rotating black hole can give up at most half of its mass–energy—no more, no less—no matter how clever the extraction."

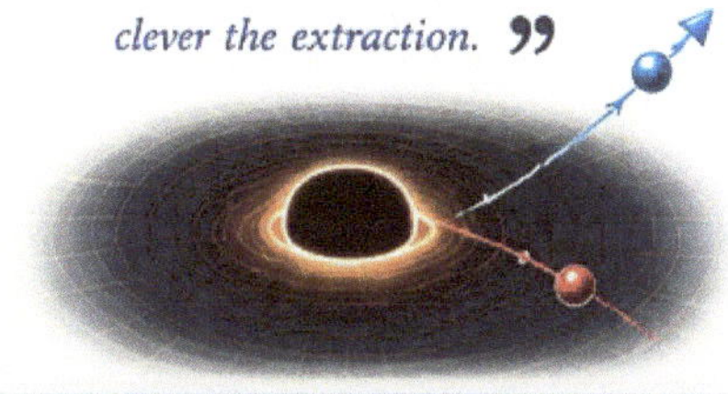

"Negative energy is not a paradox—it is nature's way of letting black holes share their hidden power."

21

SUPERRADIANCE

WAVES CAN EXTRACT ENERGY FROM BLACK HOLES

Not only particles—waves can also tap the rotational energy of a black hole.
This wave amplification is called *superradiance*.

1. The Idea

Certain wave scattering off a rotating (Kerr) black hole can come out with more energy than they had going in.

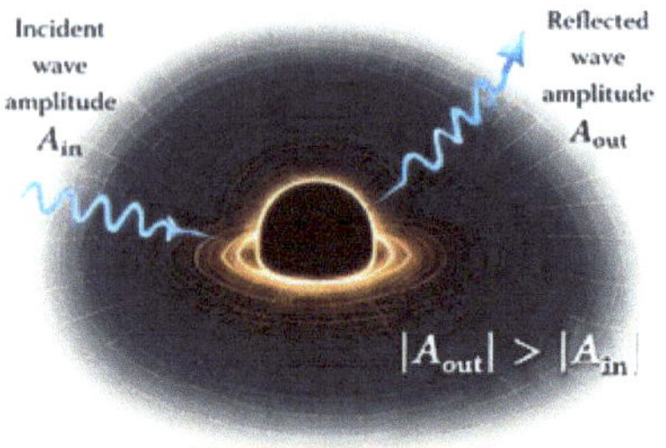

The black hole's spin does the "extra work"!

2. The Physical Picture

Inside the ergosphere, negative energy states exist. Waves can scatter so that part of the wave has negative energy and falls in.

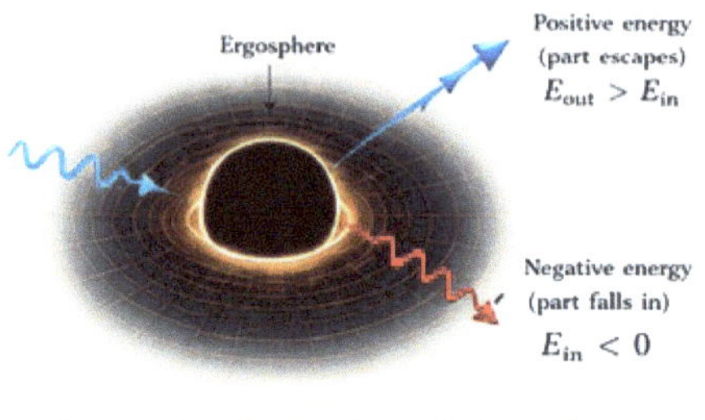

Conservation of energy is satisfied overall—the black hole loses rotational energy.

3. Superradiance Condition

Superradiance occurs only for waves with frequency ω and azimuthal number m satisfying:

$$0 < \omega < m\,\Omega_H$$

where Ω_H is the angular velocity of the horizon:

$$\Omega_H = \frac{a}{2Mr_+}$$

with horizon radius

$$r_H = M + \sqrt{M^2 - a^2}$$

$(G = c = 1$ units)

If $\omega > m\,\Omega_H$, waves are absorbed as usual. Only in the range above do they amplify.

4. Energy and Amplification

The energy flux of the reflected wave is larger than the incident flux:

$$|A_{out}|^2 > |A_{in}|^2$$

Amplification factor:

$$Z = \frac{|A_{out}|^2}{|A_{in}|^2} > 1$$

For $0 < \omega < m\,\Omega_H$, we have $Z > 1$.

Superradiance is a classical wave effect in curved spacetime.

5. Simple Example: Scalar Waves

For a scalar wave in the Kerr background $(a > 0)$, solve the wave equation $\Box\Phi = 0$.
At large r, solutions behave like:

$$\Phi \sim A_{in}\, e^{-i\omega t + im\phi} + A_{out}\, e^{-i\omega t - im\phi}$$

The reflection coefficient R satisfies:

$$R = \frac{|A_{out}|^2}{|A_{in}|^2} > 1 \qquad (0 < \omega < m\,\Omega_H)$$

Maximum amplification occurs for $\omega \lesssim m\,\Omega_H$.

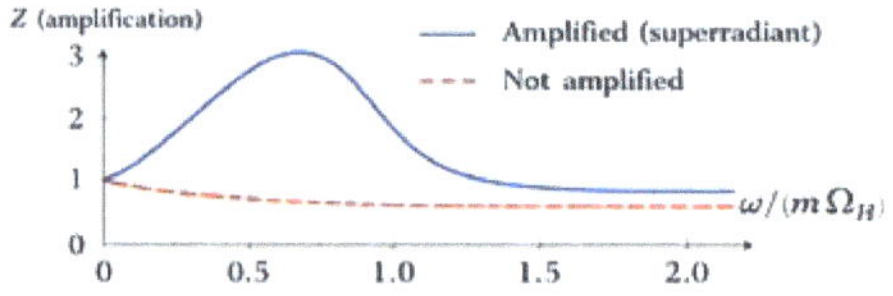

6. Fate of the Amplified Wave

The amplified wave can extract rotational energy repeatedly if a "mirror" or potential barrier traps it outside the black hole.

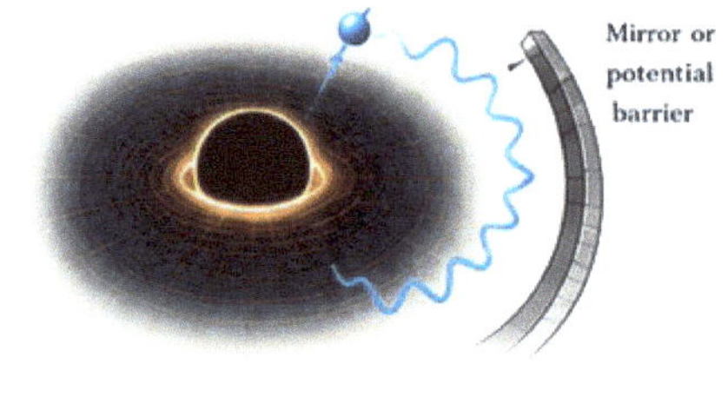

This leads to exponential growth of the wave—the black hole spins down.

7. Astrophysical Mechanisms

Superradiance can power several important astrophysical processes:

- **Black hole bombs:** A black hole + mirror (or cloud) → extraction of rotational energy.
- **Pulsar-like emission models:** Rotating black holes can emit coherent radio-like signals.
- **Gravitational wave clouds:** Axion-like particles around black holes can form clouds via superradiance.

Superradiance may explain observations of fast radio bursts, jets, and gravitational waves.

8. External: Black Hole Bomb

A mirror at radius r_m traps the wave.
Each round trip extracts energy.

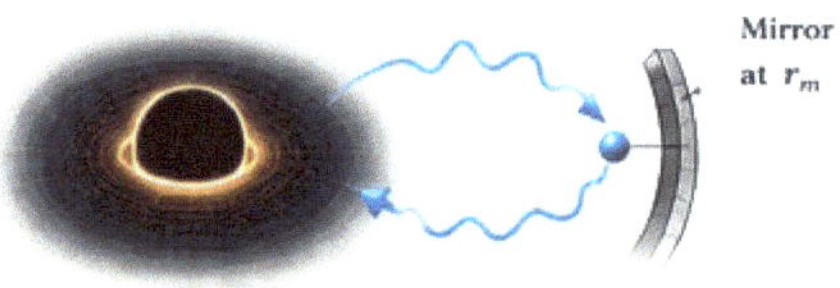

Amplification per cycle: $Z > 1$
After N cycles: $E_N = Z^N E_0$

If Z is slightly above 1, growth is exponential with time scale $\tau \sim \frac{1}{\ln Z} T_{cycle}$.

9. Superradiance vs Absorption

Waves with different frequencies behave differently.

	Superradiant Regime $(0 < \omega < m\Omega_H)$	Absorptive Regime $(\omega > m\Omega_H)$
Condition	$0 < \omega < m\Omega_H$	$(\omega > m\Omega_H)$
Outcome	$\lvert A_{out}\rvert^2 > \lvert A_{in}\rvert^2$ Energy extracted	$\lvert A_{out}\rvert^2 < \lvert A_{in}\rvert^2$ Energy absorbed
Physical meaning	Wave loses energy to the BH	Wave gives energy to the BH

The boundary $\omega = m\,\Omega_H$ is the threshold between amplification and absorption.

10. Key Equations

- Horizon angular velocity:
$$\Omega_H = \frac{a}{2Mr_+}$$
- Superradiance condition:
$$0 < \omega < m\,\Omega_H$$
- Amplification factor:
$$Z = \frac{|A_{out}|^2}{|A_{in}|^2} > 1$$
- Horizon radius:
$$r_H = M + \sqrt{M^2 - a^2}$$

Units: $G = c = 1$.

11. Key Takeaways

- Superradiance is a classical wave effect in Kerr spacetime.
- It happens only in the ergosphere.
- Waves with $0 < \omega < m\Omega_H$ amplify.
- The black hole loses rotational energy.
- With confinement, the effect can grow exponentially (black hole bomb).
- It has important astrophysical implications (jets, clouds, radio signals, GWs).
- Superradiance is one of the most elegant ways nature allows energy extraction.

Waves can be the black hole's "lever" for releasing its hidden spin.

12. In One Sentence

> *"When the conditions are right, waves can leave a rotating black hole stronger than they arrived—stealing energy from its spin in the process."*

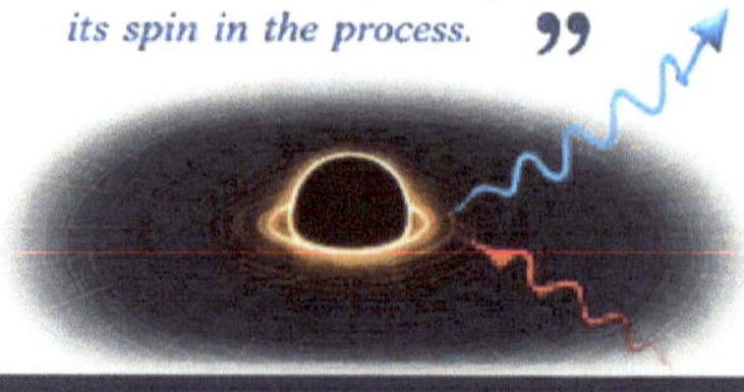

"Negative energy is not a paradox—it is nature's way of letting black holes share their hidden power."

22

Black Hole Bomb

Trapping Waves to Amplify Energy

Under the right conditions, waves can be trapped between the black hole and a reflecting boundary, growing in amplitude—like sound in an organ pipe.

1. The Basic Idea

Place a reflecting "mirror" around a rotating black hole. Waves scattered from the black hole can bounce back and forth.

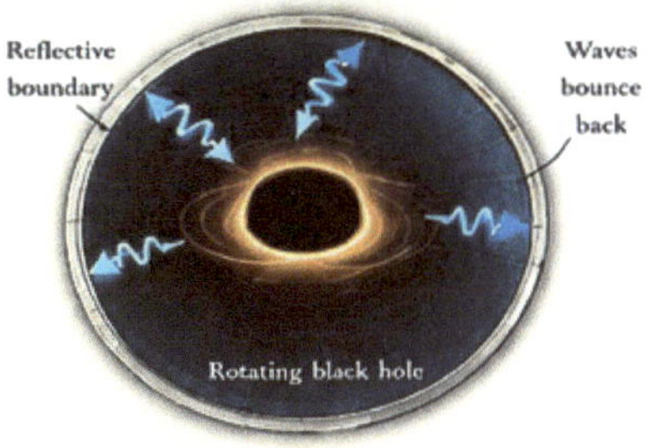

Each time the wave scatters off the black hole, it can extract a little rotational energy.

2. Superradiant Scattering

If the wave frequency satisfies

$$0 < \omega < m\Omega_H$$

reflected waves have more energy than the incident wave.

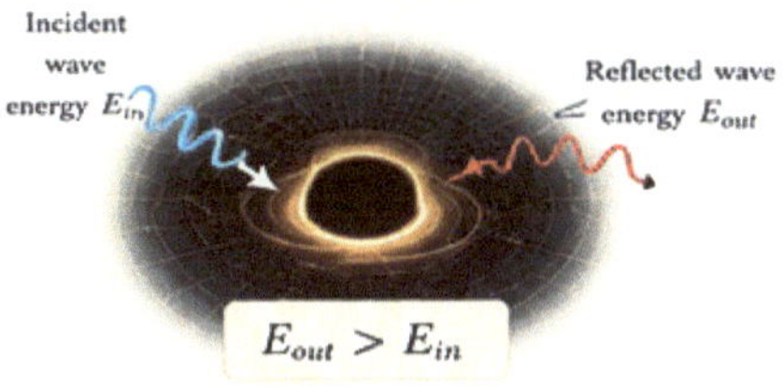

This is superradiance—energy extraction by waves.

3. The Black Hole Bomb

With a mirror outside the ergosphere, waves get amplified on each round trip.

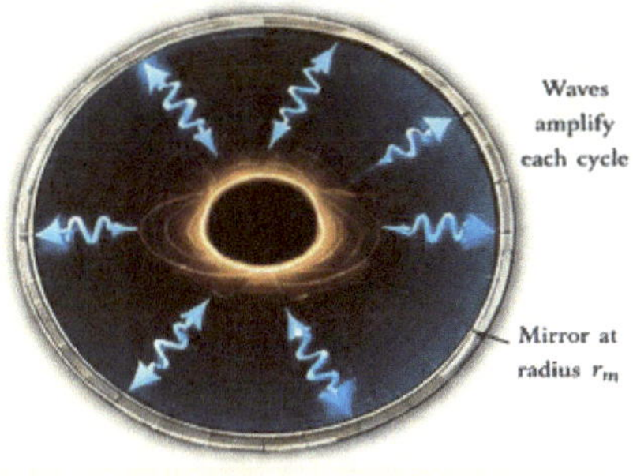

Repeated amplification leads to exponential growth of the wave's energy.

4. Condition for Instability

For a bosonic field of frequency ω and azimuthal number m in Kerr spacetime:

$$0 < \omega < m\Omega_H$$

A reflecting boundary at $r = r_m$ creates an unstable "cavity".

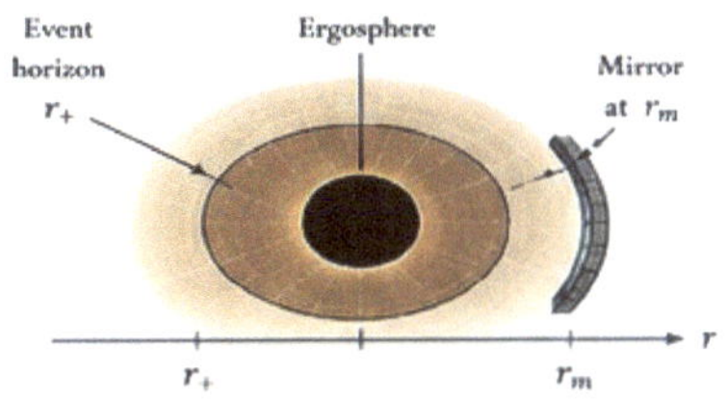

The ergosphere + mirror act like a natural resonant cavity.

5. How Amplification Works

1. Wave falls in toward the black hole.
2. Superradiant scattering: $E_{out} > E_{in}$.
3. Wave travels outward and hits the mirror.
4. It reflects and returns toward the black hole.
5. Repeat... the amplitude grows exponentially!

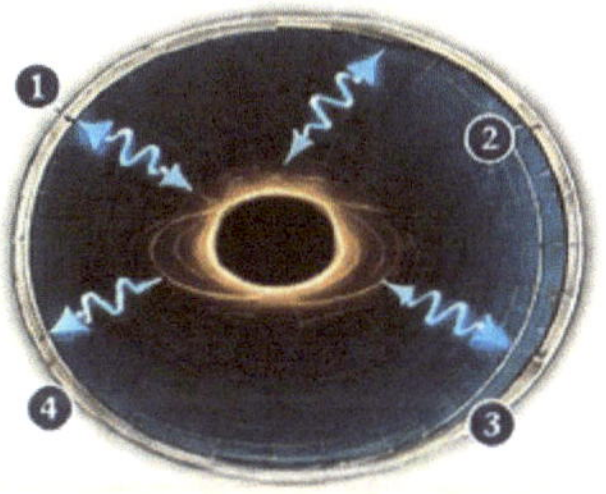

Each cycle extracts a little more rotational energy.

6. Exponential Growth

The wave amplitude grows with time:

$$|\Psi(t)| \sim e^{\Gamma t}$$

The growth rate Γ depends on:

- Black hole spin a
- Field mass μ (for massive fields)
- Mirror radius r_m
- Azimuthal number m
- Wave frequency ω

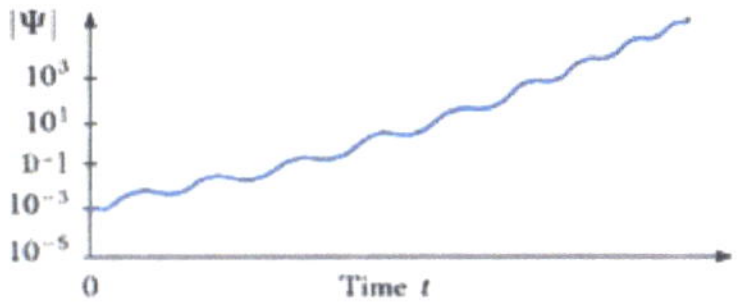

Stronger spin and the right cavity size give faster growth.

7. First Studied by Press & Teukolsky

Press & Teukolsky (1972) showed that a mirror around a Kerr black hole leads to instability.

William Press

Saul Teukolsky

A landmark discovery in black hole physics and wave theory.

8. Physical Picture

The black hole provides "gain" (superradiance). The mirror provides "feedback" (reflection). Together → a self-amplifying system.

A gravitational analog of a laser or an organ pipe!

9. Astrophysical Implications

- Could extract large amounts of energy from rotating black holes.
- May produce strong gravitational waves and electromagnetic signals.
- Motivates searches for light bosons (axions, ultralight fields).
- Could affect black hole spins in the Universe.

A black hole bomb could be nature's hidden energy machine.

10. Key Equations

- Superradiance condition:

$$0 < \omega < m\Omega_H$$

- Growth of instability:

$$|\Psi(t)| \sim e^{\Gamma t}$$

- Horizon angular velocity:

$$\Omega_H = \frac{a}{2Mr_+}$$

$$(G = c = 1)$$

11. Key Takeaways

- Waves with $0 < \omega < m\Omega_H$ are amplified.
- A mirror outside traps the wave.
- Each round trip extracts rotational energy.
- Amplitude grows exponentially.
- The system is unstable—this is the black hole bomb.
- It connects black holes, quantum fields, and observations!

Under the right conditions, black holes can power their own wave machines.

12. In One Sentence

> *"A black hole bomb shows that with the right boundary, waves can leave a rotating black hole stronger than they arrived—stealing energy from its spin in the process."*

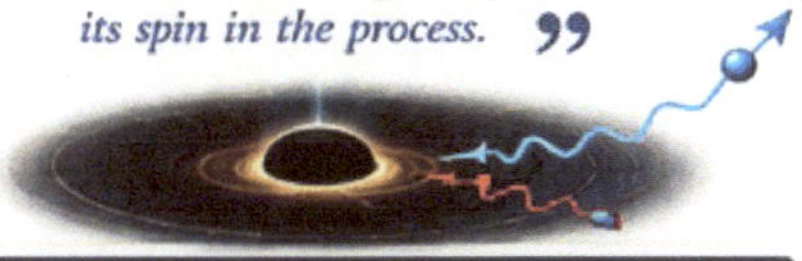

"Superradiance shows that even light and waves can unlock the rotation of a black hole."

23

Blandford–Znajek Mechanism

Electromagnetic Power Extraction

A rotating black hole threaded by magnetic fields can power jets and high-energy outflows without needing matter to enter the hole.

1. The Setup

A rotating black hole of spin a is threaded by large-scale magnetic field lines.

Magnetic field lines are anchored in a distant accretion disk or surrounding plasma.

2. Ergosphere and Field Lines

Inside the ergosphere, spacetime is dragged around by the black hole's rotation.

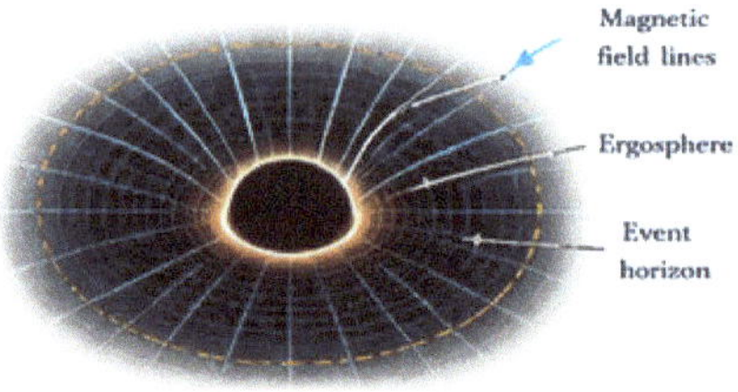

Field lines are twisted by frame dragging, storing rotational energy.

3. The Basic Idea

Twisted magnetic fields exert a torque, extracting rotational energy electromagnetically.

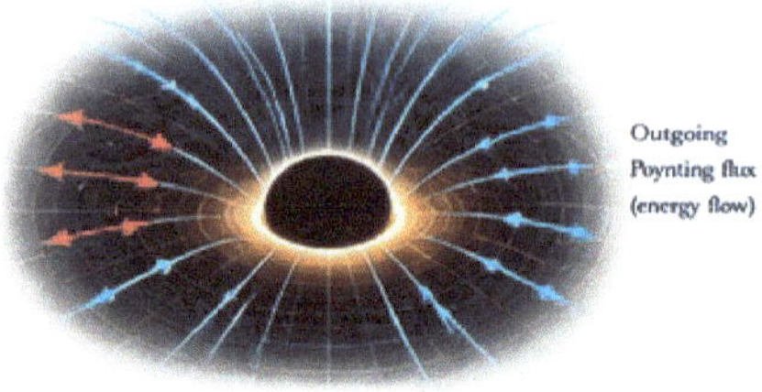

Energy is carried away by electromagnetic radiation—no mass needs to fall in.

4. Force-Free Magnetosphere

Plasma inertia is negligible: electromagnetic forces dominate.

$$J \times B = 0$$

J = electric current density
B = magnetic field

The magnetosphere is described by the force-free approximation.

5. Energy and Angular Momentum Flux

The Poynting flux (electromagnetic energy flux) escapes to infinity along the field lines.

$$S = \frac{1}{\mu_0} E \times B$$

E = electric field
B = magnetic field
μ_0 = permeability of free space

Angular momentum is carried away, so the black hole spins down.

6. BZ Power Formula (Approximate)

For a split-monopole magnetic field, the extracted power is

$$P_{\rm BZ} = \frac{\kappa}{4\pi c} \Phi_{\rm BH}^2 \Omega_H^2 f(\Omega_F/\Omega_H)$$

where

Φ_{BH} = magnetic flux threading the horizon
Ω_H = angular velocity of the horizon
Ω_F = angular velocity of field lines
κ ≈ 0.05 (dimensionless constant)
$f(x) = x^2(1-x)^2,\ 0 < x < 1$

Maximum power at $\Omega_F \approx \Omega_H / 2$.

7. Horizon Angular Velocity

The horizon angular velocity is

$$\Omega_H = \frac{ac}{2Mr_+}$$

with

$$r_+ = M + \sqrt{M^2 - a^2}$$

$(G = c = 1)$

where M is the black hole mass.

Faster spin $(a \to 1)$ gives larger Ω_H and more extractable power.

8. Circuit Analogy (Membrane View)

The black hole acts like a battery, with EMT driving current through the external circuit.

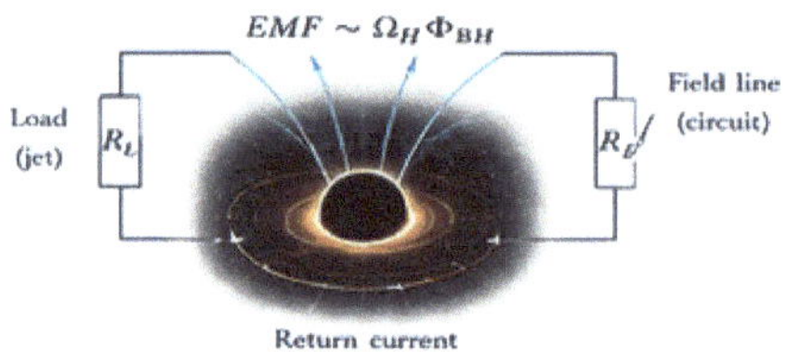

Power is delivered to the load (e.g., jets) through electromagnetic currents.

9. Astrophysical Jets

BZ power can launch and sustain relativistic jets in AGN and gamma-ray bursts.

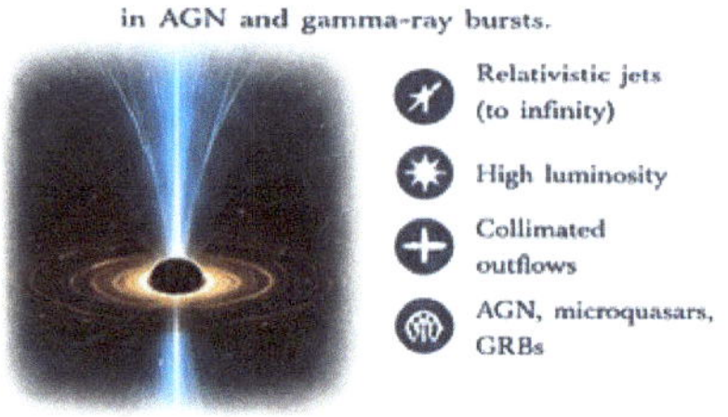

Magnetic fields tap the black hole's spin to power some of the Universe's most energetic jets.

10. Conditions for Efficiency

- Strong magnetic flux: large $\Phi_{\rm BH}$
- High spin: $a \lesssim 1$
- Force-free magnetosphere $\Omega_H = \frac{a}{2Mr_+}$
- BZ power (approx.): $P_{\rm BZ} \propto \Phi_{\rm BH}^2 \Omega_H^2$

Optimal field line rotation:

Open field lines are essential—closed loops do not extract net energy.

11. BZ vs. Matter-Based Extraction

Feature	BZ Mechanism	Penrose Process	Accretion Power
Medium	Electromagnetic field	Particles	Matter accretion
Requires matter inflow?	No	Yes	Yes
Efficiency limit	High ($a \to 1$)	$\leq 21\%$	Up to ~ 0.4
Energy carrier	Poynting flux	Particles	Radiation + jets
Astrophysical role	Powers jets	Minor	Luminosity, disks

BZ is a clean, electromagnetic way to extract rotational energy.

12. Numerical Simulations

GRMHD simulations confirm the BZ mechanism. Jet power grows with magnetic flux and spin.

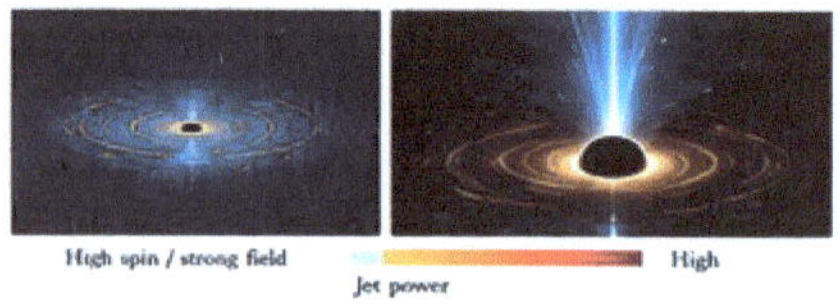

Stronger fields and higher spin produce more powerful jets.

13. Key Equations Summary

- Horizon radius: $r_+ = M + \sqrt{M^2 - a^2}$
- Horizon angular velocity: $\Omega_H = \frac{a}{2Mr_+}$
- BZ power (approx.): $P_{\rm BZ} \propto \Phi_{\rm BH}^2 \Omega_H^2$
- Optimal field line rotation: $\Omega_F \approx \Omega_H / 2$

14. Key Takeaways

- Rotating black holes threaded by magnetic fields can lose energy electromagnetically.
- Twisted field lines in the ergosphere generate a Poynting flux that escapes to infinity.
- No matter needs to cross the horizon—energy is extracted through fields.
- The power increases with magnetic flux and black hole spin.
- The Blandford–Znajek mechanism explains many powerful astrophysical jets.

The black hole becomes a cosmic dynamo. Its spin energy lights up the Universe.

15. In One Sentence

"Magnetic fields can tap the spin of a black hole and launch jets by carrying energy away as electromagnetic waves, without needing matter to fall in."

"A spinning black hole plus magnetic fields—and the black hole keeps giving energy away."

Astrophysical Jets

Collimated Outflows Powered by Black Holes

The Blandford–Znajek mechanism can launch powerful, narrow jets of plasma from the poles of a rotating black hole, impacting galaxies and the cosmos.

1. The Phenomenon

Many black holes launch narrow, high-speed jets along their rotation axis.

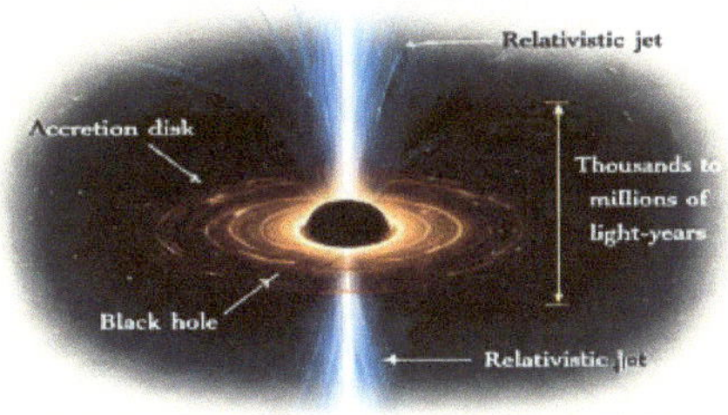

Jets are observed in quasars, blazars, AGN, microquasars, and some gamma-ray bursts.

2. Origin of the Jet

Rotating magnetic fields extract energy from the black hole and launch plasma along field lines.

Energy flows outward along open field lines anchored in the accretion disk.

3. Polar Launching

Jets are launched from the polar regions where magnetic field lines are open to infinity.

Only open field lines allow Poynting flux to escape and form a jet.

4. Energy and Power

BZ power increases with magnetic flux and black hole spin.

$$P_{BZ} = \frac{\kappa}{4\pi c}\,\Phi_{BH}^2\,\Omega_H^2\,f(\Omega_F/\Omega_H)$$

Where:

- Φ_{BH} = magnetic flux threading the horizon
- Ω_H = angular velocity of the horizon
- Ω_F = angular velocity of field lines
- $\kappa = \sim 0.05$ (dimensionless constant)
- $f(x) = x^2\,(1-x)^2,\ 0 < x < 1$

Maximum power at $\Omega_F \approx \Omega_H/2$.

5. Jet Structure

Astrophysical jets are highly collimated and remain stable over enormous distances.

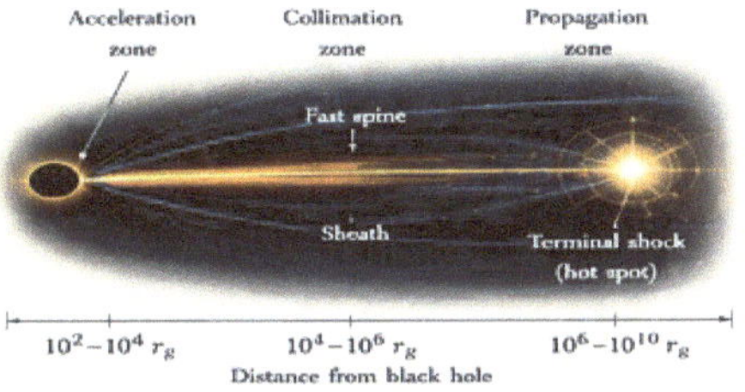

Jets have a fast inner spine surrounded by a slower sheath and end in a hot spot.

6. Observational Evidence

Jets are seen across a wide range of scales.

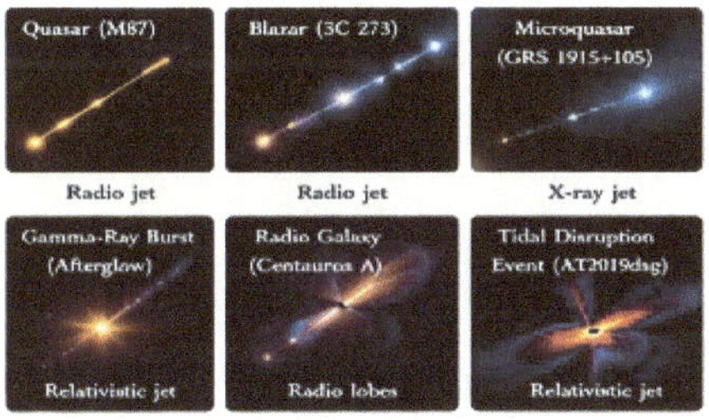

From stellar-mass to supermassive black holes, jets are a universal phenomenon.

7. Jet Collimation

Magnetic hoop stress collimates the jet into a narrow beam.

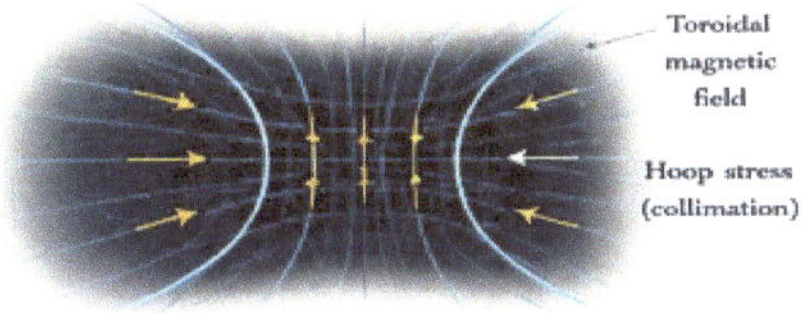

The toroidal magnetic field creates inward stress that keeps the jet narrow and stable.

8. Feedback in Galaxies

Jets deposit energy into the interstellar and intergalactic medium.

Jets can heat gas, suppress star formation, and regulate galaxy evolution.

9. Connection to Gamma-Ray Bursts

In collapsars, a newly formed black hole can launch a relativistic jet.

The jet breaks out of the star and produces a gamma-ray burst when viewed along the jet.

10. Jet Composition

Jets may contain electrons, positrons, and protons.

Component	Role	Evidence
Electrons / Positrons	Radiation (synchrotron, IC scattering)	X-ray, optical, gamma-ray emission
Protons	Carry kinetic energy, heat the medium	Energy transport, shock heating
Magnetic Fields	Provide structure, collimation, and energy transport	Polarization, stability

The exact composition remains an open question.

11. Key Numbers

- Lorentz factor: $\Gamma \sim 10 - 50$ (typical)
- Jet power: $10^{44} - 10^{47}$ erg s^{-1} (AGN)
- Magnetic field near BH: $10^3 - 10^5$ G
- Scale: $10^2 - 10^{10}$ gravitational radii
- Lifetime: $10^6 - 10^8$ years

12. Simulation Snapshot

GRMHD simulations reproduce jet launching.

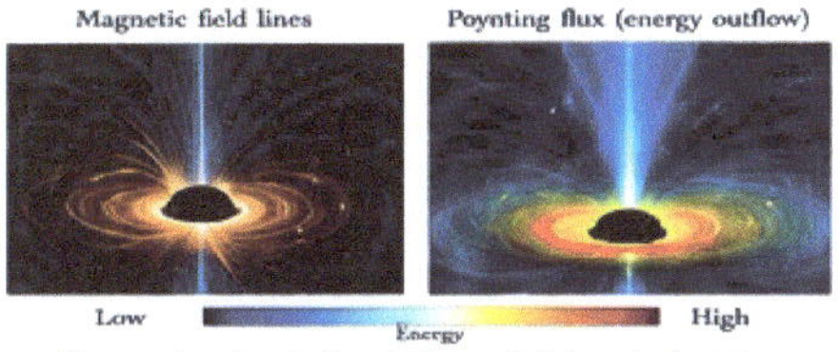

Energy is extracted and channeled into twin jets.

13. Key Equations Summary

- BZ power (approx.):

$$P_{BZ} \propto \Phi_{BH}^2\,\Omega_H^2\,f(\Omega_F/\Omega_H)$$

- Horizon angular velocity:

$$\Omega_H = \frac{a}{2Mr_+}$$

- Optimal field line rotation:

$$\Omega_F \approx \frac{\Omega_H}{2}$$

- Horizon radius:

$$r_+ = M + \sqrt{M^2 - a^2}$$

$$(G = c = 1)$$

14. Key Takeaways

- Rotating black holes threaded by magnetic fields can launch powerful, collimated jets.
- Twisted field lines in the ergosphere generate a Poynting flux that escapes to infinity.
- No matter needs to cross the horizon—energy is extracted through fields.
- The power increases with magnetic flux and black hole spin.
- The Blandford–Znajek mechanism explains many powerful astrophysical jets.

The black hole becomes a cosmic dynamo. Its spin energy lights up the Universe.

15. In one Sentence

“Through the Blandford–Znajek mechanism, a rotating black hole can launch magnetic jets of energy that shape galaxies and illuminate the Universe.”

“Black holes do not just consume—they can also give, launch jets and energy that shape the cosmos.”

25

Three Energy Carriers

How Black Holes Deliver Energy to the Universe

Black holes can transfer energy to the outside world in different ways.
The Blandford–Znajek mechanism is one of three main energy carriers.

1. Overview

Energy can be extracted from a rotating black hole and carried away by three channels.

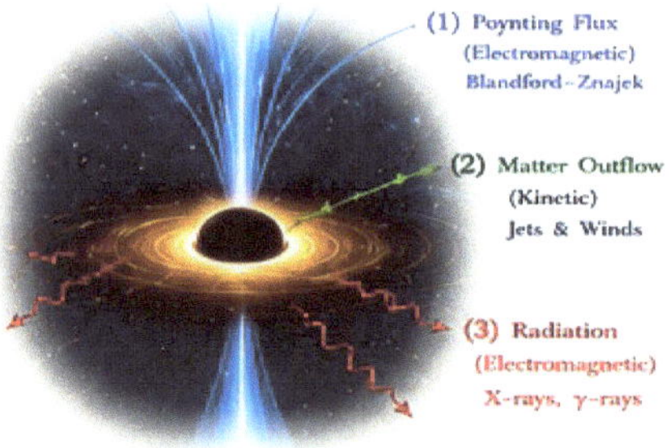

These channels operate together and power some of the Universe's most energetic phenomena.

2. The Three Channels

Energy leaves the black hole through electromagnetic fields, matter, or radiation.

1. **Poynting Flux** (*Blandford–Znajek*)
Electromagnetic energy is carried away by large-scale magnetic fields that thread the horizon.

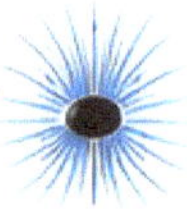

2. **Matter Outflow** (Jets & Winds)
Plasma is accelerated and ejected along the rotation axis, carrying kinetic energy.

3. **Radiation** (Electromagnetic)
Energy is released as radiation across the spectrum—radio, X-rays, optical, γ-rays.

All three are observed in many black hole systems, from stellar-mass to supermassive scales.

3. Blandford–Znajek Power

A rotating black hole threaded by magnetic flux can launch Poynting flux.

$$P_{BZ} = \frac{\kappa}{4\pi c}\, \Phi_{BH}^2 \Omega_H^2\, f(\Omega_F/\Omega_H)$$

where:

- Φ_{BH} = magnetic flux threading the horizon
- Ω_H = angular velocity of the horizon
- Ω_F = angular velocity of field lines
- $\kappa \sim 0.05$ (dimensionless constant)
- $f(x) = x^2(1-x)^2,\ 0 \le x \le 1$

Maximum power at $\Omega_F = \Omega_H/2$

This is the most efficient electromagnetic energy extraction process known.

4. Matter Outflows

Jets are launched when plasma is accelerated by electromagnetic and thermal forces.

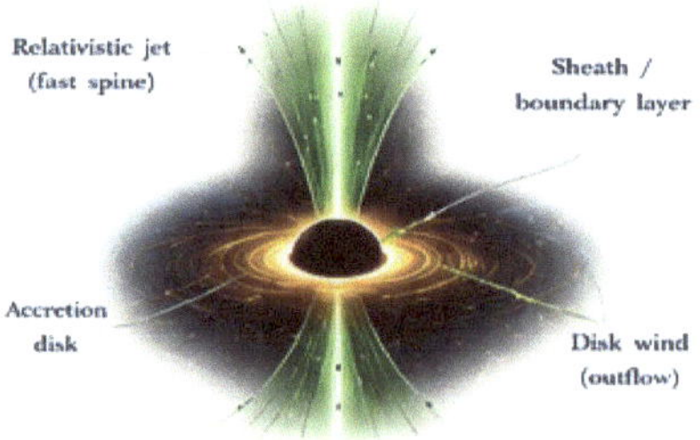

Jets carry kinetic energy of matter, which can be converted into radiation far from the black hole.

5. Radiation Channels

Particles in jets and disks radiate across the electromagnetic spectrum.

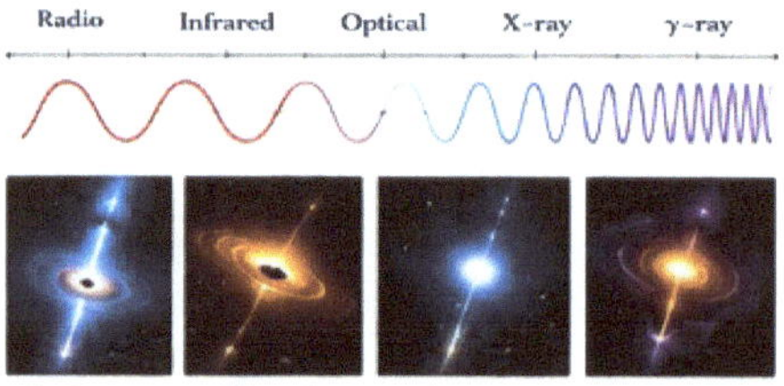

Synchrotron (radio–optical) | Thermal (disk) | Compton (X-rays) | Inverse Compton / Hadronic (γ-rays)

Radiation is the channel we detect most easily, revealing the presence of black holes.

6. Comparing the Channels

Channel	Carrier	Typical Scale	Efficiency (Relative)	Where Seen
1 Poynting Flux (BZ)	Electromagnetic fields	Near horizon to large scales	High	AGN jets, GRBs, microquasars
2 Matter Outflow	Plasma (kinetic)	pc to Mpc	Medium	Jets, winds, radio lobes
3 Radiation	Photons	All scales	Varies	All black hole systems

Poynting flux is often dominant close to the black hole, but all channels work together.

7. How They Work Together

Magnetic fields extract energy (BZ), accelerate matter (jets), and drive shocks that produce radiation.

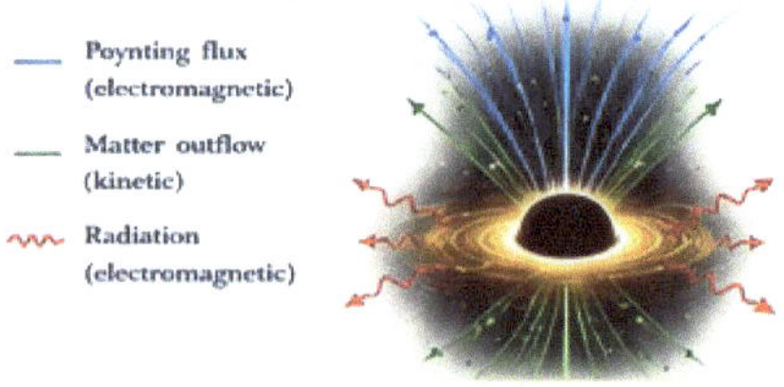

The three channels are interconnected parts of one powerful engine.

8. Examples Across the Universe

Steady jets, radio & X-rays | Powerful jets, radio lobes, broad spectrum | Ultra-relativistic jets, γ-ray emission | X-ray/optical flares, outflows

From nearby stellar systems to the farthest quasars, the same three energy carriers appear.

9. Observational Signatures

Radio Lobes
Traces of past jet activity inflating bubbles in intergalactic space.

Relativistic Jets
Narrow beams moving close to the speed of light, often seen in AGN and GRBs.

High-energy Radiation
X-rays and γ-rays reveal extreme accelerated particles and shocks.

Polarization
Synchrotron emission is polarized, confirming magnetic field roles.

Observations confirm that black holes are nature's most powerful energy factories.

10. Key Equations

- BZ Power (maximum):

$$P_{BZ,max} = \frac{\kappa}{16\pi c}\, \Phi_{BH}^2 \Omega_H^2$$

- Horizon Angular Velocity:

$$\Omega_H = \frac{a}{2Mr_+}$$

- Horizon Radius:

$$r_+ = M + \sqrt{M^2 - a^2}$$

$$(G = c = 1)$$

11. Key Takeaways

- ✔ Black holes can send energy out in three main ways: magnetic fields, matter, and radiation.
- ✔ The Blandford–Znajek mechanism taps rotation energy via magnetic fields.
- ✔ Jets carry kinetic energy that shapes galaxies and clusters.
- ✔ Radiation across the spectrum reveals the physical processes at work.
- ✔ Together, these channels connect black holes to the evolution of the Universe.

The black hole becomes a cosmic dynamo.
Its spin energy lights up the Universe.

12. In One Sentence

> *"Black holes are not silent absorbers—they are engines that send out magnetic power, matter, and light to shape the cosmos."*

"Three energy carriers, one cosmic engine—black holes power the Universe."

One Energy Reservoir

The Unified View of Black Hole Energy

All the processes we have seen—waves, fields, particles, and jets—are different expressions of the same engine: the rotational energy of a black hole.

1. The Central Fact

A rotating black hole contains enormous rotational energy in its spin.

$$E_{rot} = Mc^2 \left(1 - \frac{\sqrt{1-a^2}}{2}\right)$$

where $a = \frac{J}{M^2}$ is the dimensionless spin.

This rotational energy is the ultimate source for all energy extraction mechanisms.

2. Different Paths, Same Source

Three unique ways the black hole through its rotation powers three main energy carriers.

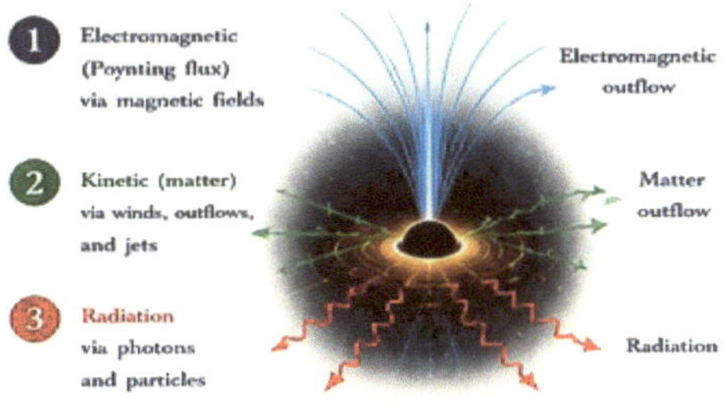

No matter how it leaves, the energy originates from spacetime rotation.

3. The Big Picture

Imagine the black hole as a cosmic power plant. Spin is the fuel, and these three carriers are the power lines.

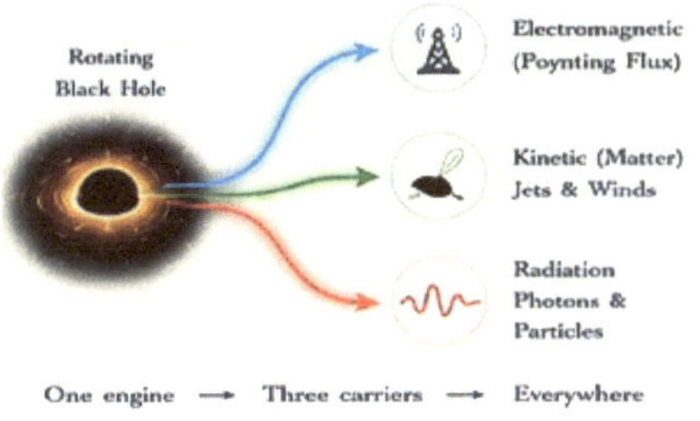

This unified view explains the richness of black hole phenomena across the Universe.

4. Energy Flow Diagram

Energy extracted from the ergosphere is carried away by different channels.

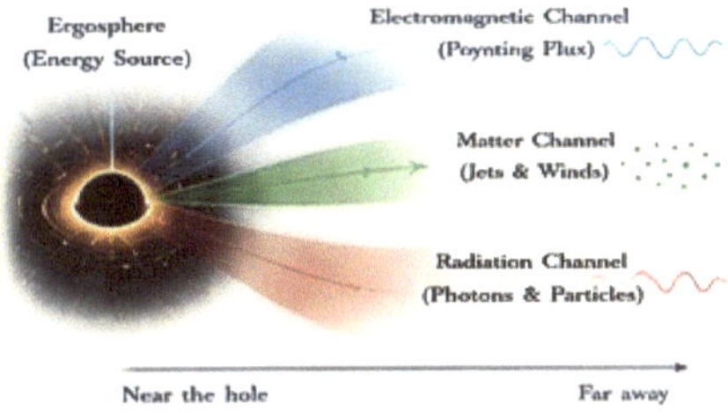

Each channel carries energy outward and leaves signatures we can observe.

5. Where We See Them

These energy carriers are observed across a wide range of black hole systems.

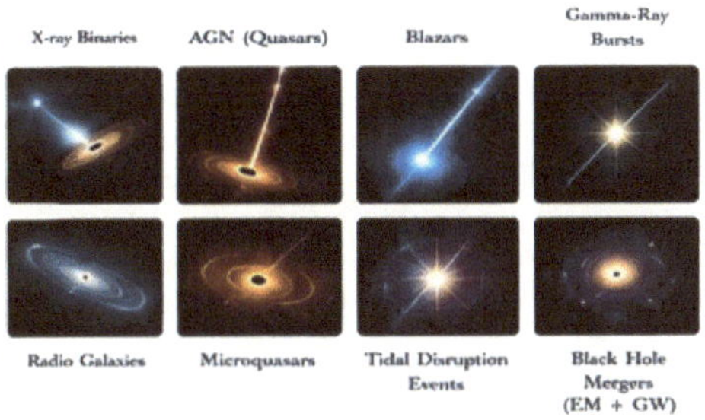

From stellar to cosmic scales, nature uses the same engine with different power outputs.

6. How the Energy is Set Free

Several physical mechanisms tap into the spin.

- **Blandford–Znajek (Electromagnetic)** Large-scale magnetic fields threading the horizon extract energy electromagnetically.
- **Blandford–Payne (Matter Outflow)** Magnetized disk winds launch matter outward along open field lines.
- **Penrose Process (Particles/Waves)** Particles and waves gain energy by splitting inside the ergosphere.
- **Hawking Radiation (Quantum)** Quantum effects release radiation, dominant for small black holes.

Different mechanisms, same fuel.
The more spin and magnetic flux, the more power.

7. Efficiency Comparison

Different carriers have different efficiencies depending on conditions.

Carrier	Mechanism	Efficiency (Typical)	Best For
Electromagnetic (Poynting Flux)	Blandford–Znajek	~1% – 300%*	Powering relativistic jets
Matter (Jets, Winds)	Blandford–Payne	~0.1% – 10%	Launching outflows
Radiation (Photons)	Penrose, Synchrotron, Compton, etc.	~0.01% – 5%	Emission at all scales
Hawking (Quantum)	Quantum effects	Tiny (tiny mass BHs)	Evaporation of micro BHs

*Can reach > 300% in ideal force-free, high-spin cases.

Electromagnetic extraction is the most efficient for powering the most extreme jets.

8. The Engine Analogy

Just like a car engine can drive different tools, a black hole's spin can power many phenomena.

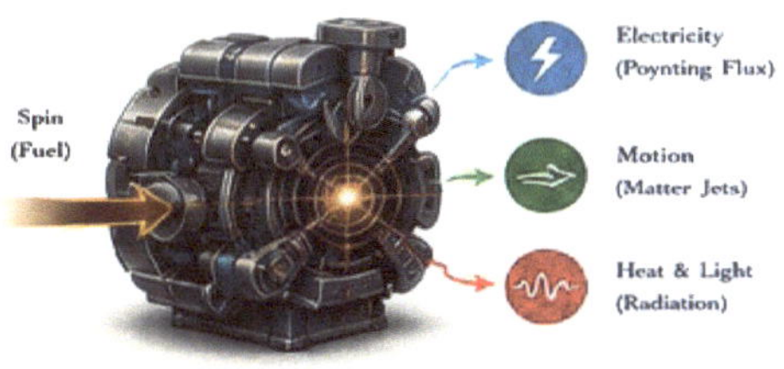

One fuel—many outputs.
Black holes are nature's most powerful engines.

9. Observational Signatures

Each energy carrier leaves a unique fingerprint.

- **Electromagnetic** Radio lobes, jets, high-energy gamma rays, polarization.
- **Matter** Broad emission lines, winds, jet knots, kinetic feedback.
- **Radiation** Thermal emission, X-rays, synchrotron, inverse Compton.
- **Quantum** Hawking spectrum (very weak), black hole evaporation (early Universe?).

By reading the signatures, we learn how the engine is working.

10. Unified Equation (Max Power)

The maximum extractable power (BZ mechanism) for an optimal magnetic field configuration is

$$P_{BZ,max} = \frac{\kappa}{4\pi c} \Phi_{BH}^2 \Omega_H^2$$

where:

- Φ_{BH} = magnetic flux threading the horizon
- Ω_H = angular velocity of the horizon
- $\kappa \sim 0.05$ (dimensionless constant)

$$P_{max} \propto B^2 M^2 a^2$$

More spin, stronger fields, bigger black hole—more power.

11. Key Takeaways

- Black holes can send energy out in three main ways: magnetic fields, matter, and radiation.
- The Blandford–Znajek mechanism taps rotational energy via magnetic fields.
- Jets carry kinetic energy that shapes galaxies and clusters.
- Radiation across the spectrum reveals the physical processes at work.
- Together, these channels connect black holes to the evolution of the Universe.

The black hole becomes a cosmic dynamo.
Its spin energy lights up the Universe.

12. In One Sentence

> *"Black holes are not just cosmic vacuum cleaners—they are energy engines, using their spin to power jets, light, and motion across the Universe."*

"Three energy carriers, one cosmic engine—black holes power the Universe."

27

Energy Flow around a Kerr Black Hole

The Overall Picture — Regions, Flows, and Extraction Pathways

All energy extraction mechanisms—Penrose process, superradiance, Blandford–Znajek jets, and Hawking radiation—arise from the same spacetime structure and energy flows.

1. The Kerr Black Hole

A rotating black hole has mass M and spin a $(0 \le a < M)$.

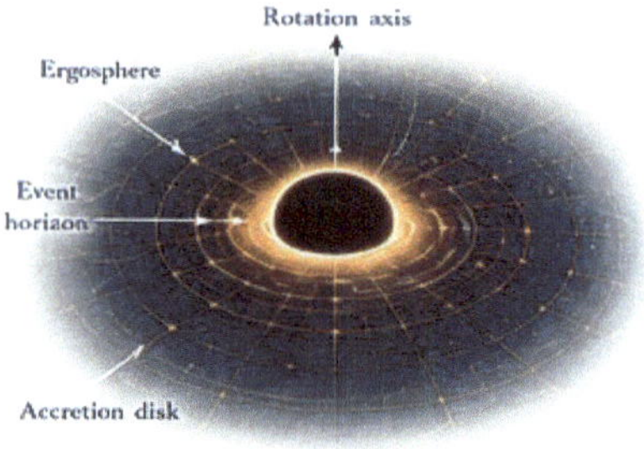

Outside the event horizon lies the ergosphere, where no object can remain at rest. It is the engine room of energy extraction.

2. Regions of Spacetime

The Kerr spacetime has distinct regions that govern energy behavior.

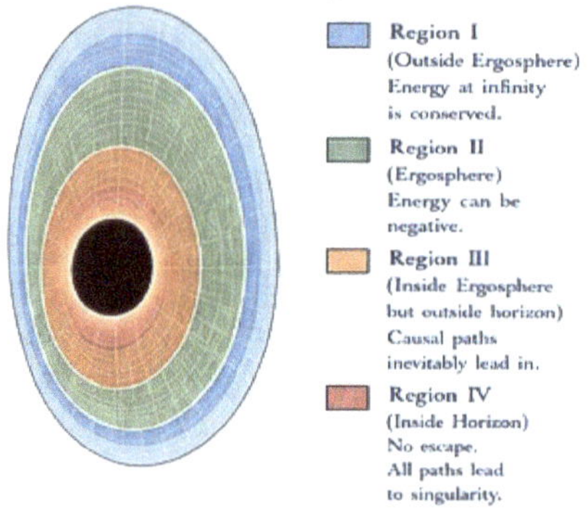

The ergosphere (Region II) is the gateway where energy extraction becomes possible.

3. Negative Energy Orbits

Inside the ergosphere, frame dragging is so strong that negative energy states ($E < 0$) are allowed.

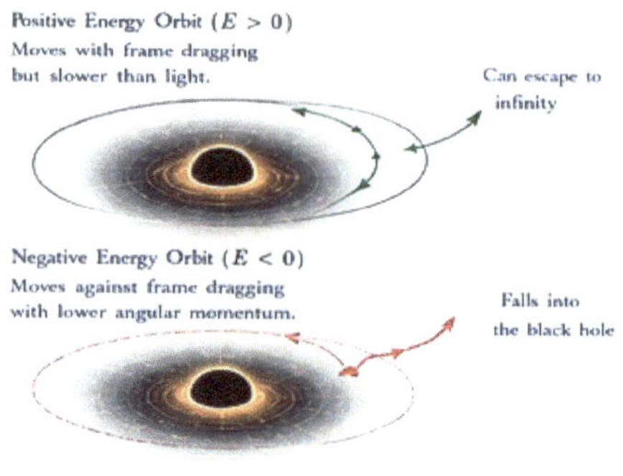

Negative energy orbits are the key to powering the universe.

4. Energy Flow Map

Energy flows in the ergosphere form closed pathways for extraction.

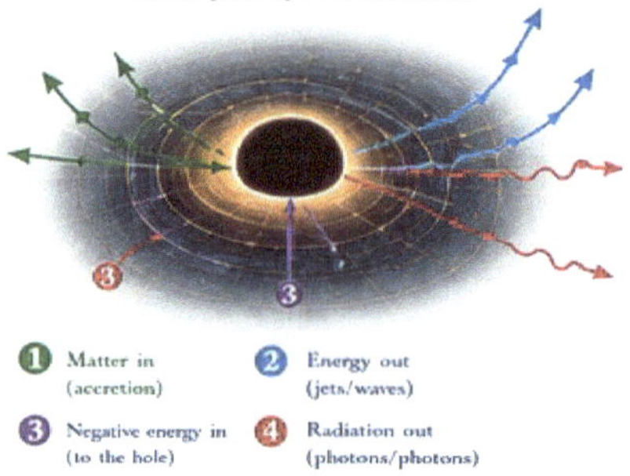

All extraction mechanisms follow this unified energy flow cycle.

5. The Four Extraction Mechanisms

Different faces of the same energy engine.

1. Penrose Process (Particles)
A particle splits inside the ergosphere. One fragment escapes with $E_{out} > E_{in}$, the other falls in with $E_{neg} < 0$.

2. Superradiance (Waves)
Waves with $0 < \omega < m\Omega_H$ are amplified. The wave extracts rotational energy.

3. Blandford–Znajek Mechanism (Fields)
Magnetic field lines threading the horizon launch Poynting flux jets.

4. Hawking Radiation (Quantum)
Quantum vacuum fluctuations produce pairs; one escapes, one falls in.

Particles, waves, fields, and quantum effects—all extract energy by the same principle.

6. Unified Energy Cycle

This cycle powers jets, radiation, and the evolution of black holes.

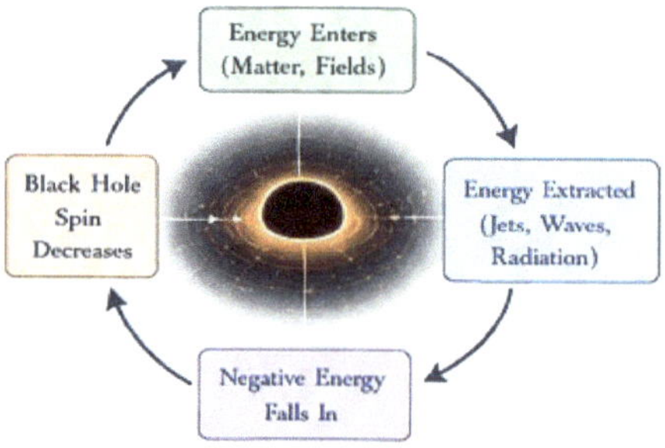

The black hole loses spin, extracts energy, and drives the cosmos.

7. Effect on Black Hole

Energy extraction reduces the black hole's spin and mass over time.

Mechanism	Effect on Black Hole	
	Spin	Mass/Energy
Penrose Process	Reduces spin	Loses mass ΔM
Superradiance	Reduces spin	Loses energy in waves
Blandford–Znajek	Reduces spin	Loses mass in jets
Hawking Radiation	Reduces mass M	(spin also slowly)

Over cosmic time, the black hole evolves toward a slower, lighter state.

8. The Engine Analogy

Astrophysical systems where energy extraction powers cosmic phenomena.

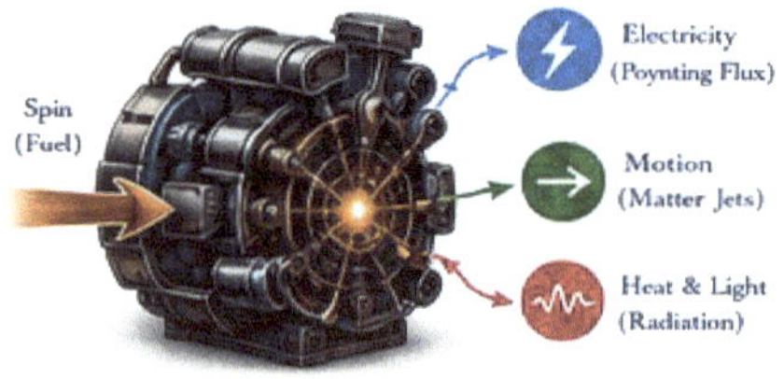

One fuel—many outputs.
Black holes are nature's most powerful engines.

9. Observational Signatures

Each energy carrier leaves a unique fingerprint.

- **Electromagnetic** — Radio lobes, jets, high-energy gamma rays, polarization.
- **Matter** — Broad emission lines, winds, jet knots, kinetic feedback.
- **Radiation** — Thermal emission, X-rays, synchrotron, inverse Compton.
- **Quantum** — Hawking spectrum (very weak), black hole evaporation (early Universe?).

By reading the signatures, we learn how the engine is working.

10. Key Equations Summary

- BZ Power (maximum):

$$P_{BZ,\max} = \frac{\kappa}{4\pi c}\,\Phi_{BH}^2\,\Omega_H^2$$

- Horizon Angular Velocity:

$$\Omega_H = \frac{ac}{2Mr_+}$$

- Horizon Radius:

$$r_+ = M + \sqrt{M^2 - a^2}$$

$$(G = c = 1)$$

More spin, stronger fields, bigger black hole—more power.

11. Key Takeaways

- Black holes can send energy out in three main ways: magnetic fields, matter, and radiation.
- The Blandford–Znajek mechanism taps rotational energy via magnetic fields.
- Jets carry kinetic energy that shapes galaxies and clusters.
- Radiation across the spectrum reveals the physical processes at work.
- Together, these channels connect black holes to the evolution of the Universe.

The black hole becomes a cosmic dynamo.
Its spin energy lights up the Universe.

12. In one Sentence

> *"Black holes are not just cosmic vacuum cleaners—they are energy engines, using their spin to power jets, light, and motion across the Universe."*

"One reservoir of spin. Four faces of power. Infinite ways to light the Universe."

28 CLASSICAL VS QUANTUM EXTRACTION

TWO REALMS — SAME ENGINE, DIFFERENT TOOLS

Black holes release energy through electromangetic and fluid processes we understand classically, and through quantum processes like Hawking radiation. Two complementarypicturesof one reality.

1. Two Ways to Extract Energy

Black holes can release energy in classical (macroscopic) and quantum (microscopic) ways.

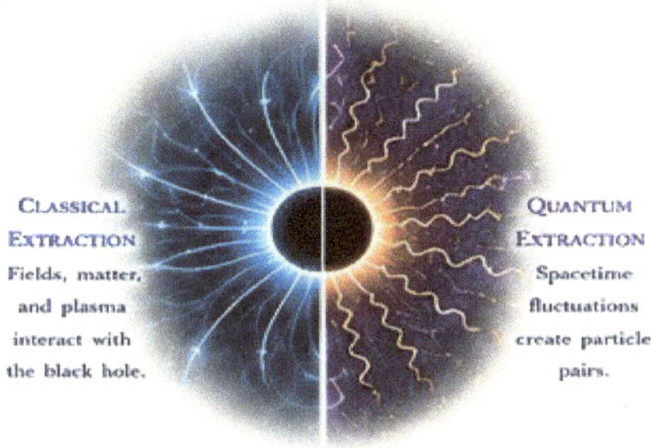

Different physics, same source: the curvature and rotation of spacetime.

2. Classical Extraction Mechanisms

Macroscopic processes tap rotational energy through fields and matter.

Blandford–Znajek (BZ) Mechanism

Magnetic field lines anchored in the disk extract energy electromagnetically.

$$P_{BZ} \approx \kappa\ \Phi_{BH}^2\ \Omega_H^2\ f(\Omega_F/\Omega_H)$$

Penrose Process

Particles enter the ergosphere and split. One falls in, the other escapes with more energy.

Accretion Power

Gravitational energy of infalling matter is converted to heat, radiation, and outflows.

These are classical, large-scale processes described by general relativity and MHD.

3. Quantum Extraction: Hawking Radiation

Quantum fluctuations near the horizon create particle–antiparticle pairs.

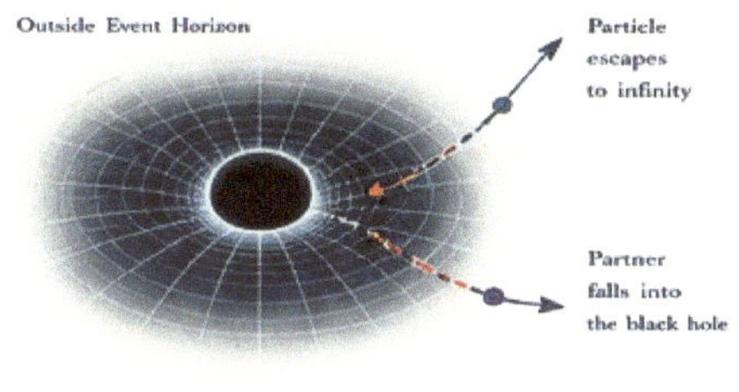

Temperature:

$$T_H = \frac{\hbar c^3}{8\pi G M k_B}$$

Power (approx.):

$$P_H \approx \frac{\hbar c^6}{G^2 M^2}$$

A quantum effect — created near the horizon. Dominant only for small (evaporating) black holes.

4. Scales and Dominance

Which mechanism dominates depends on mass, spin, magnetic field, and environment.

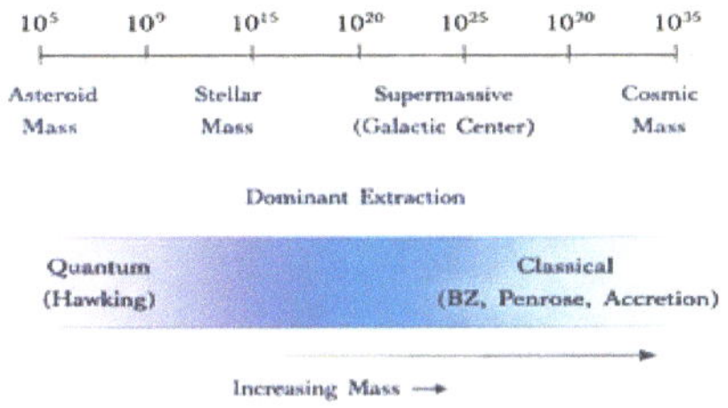

Quantum dominates only at very small masses. Classical dominates for astrophysical black holes.

5. Classical vs Quantum — Key Differences

	Classical Extraction	Quantum Extraction
Nature	Deterministic, macroscopic	Probabilistic, microscopic
Requirement	Spin, field, or matter	None (vacuum fluctuations)
Carrier	Fields, particles, plasma	Particles (photons, neutrinos, etc.)
Dominant Scale	Large black holes ($M \gtrsim 10^5\,M_\odot$)	Small black holes ($M \lesssim 10^5\,M_\odot$)
Power Dependence	$P \propto a^2 B^2 M^2 \Omega^2$	$P \propto 1/M^2$
Output	Jets, winds, radiation	Thermal spectrum (blackbody)

Two regimes, two toolkits—both describe the same black hole engine.

6. Energy Spectra

Classical processes produce non-thermal spectra. Hawking radiation is thermal.

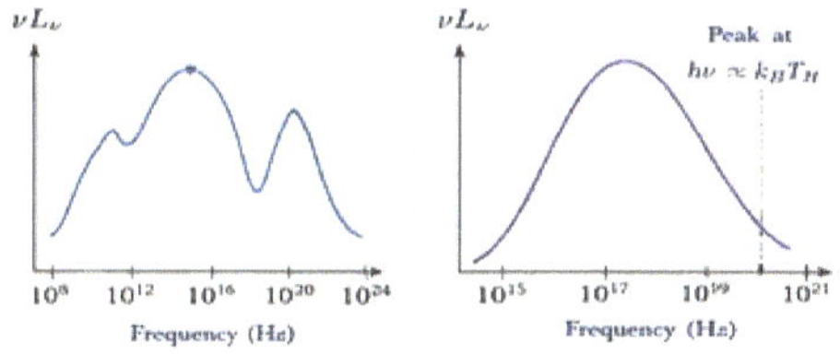

Classical = broadband, non-thermal.
Quantum = thermal, determined by T_H.

7. Environments and Conditions

The surrounding medium which processes can operate efficiently.

Favorable for Classical

- Strong magnetic fields
- Rapid spin
- Accretion disk present
- Dense plasma

Favorable for Quantum

- Isolated black hole
- No accretion disk
- Very low accretion
- Small mass

Astrophysical black holes: classical dominated.
Primordial or microscopic holes: quantum dominated.

8. Complementary, Not Competing

Both classical and quantum processes can operate at the same time.

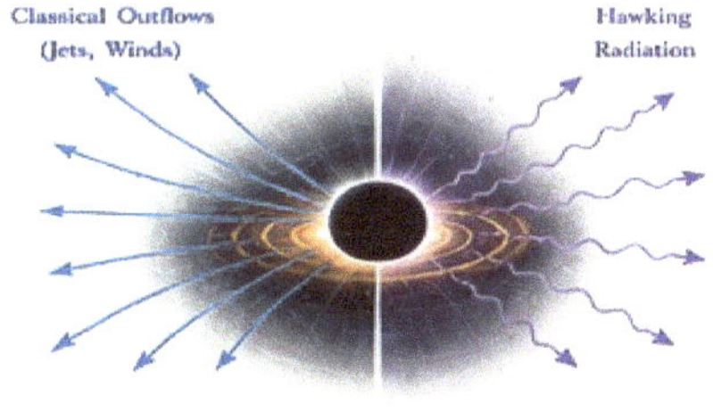

Classical processes dominate the power budget; quantum sets the ultimate fate.

9. Observational Signatures

We can distinguish the two regimes by their observable fingerprints.

- Relativistic jets
- Polarized emission
- Variability (MHD instabilities)
- Non-thermal spectra
- Large-scale lobes and bubbles

- Thermal spectrum
- No polarization
- Steady, isotropic emission
- Evaporation over time
- Possible in γ-ray background

Different fingerprints → different physics.

10. Key Equations Summary

- Classical (Maximum Power, BZ)

$$P_{BZ} = \frac{\kappa}{4\pi c}\ \Phi_{BH}^2\ \Omega_H^2\ f(\Omega_F/\Omega_H)$$

- Horizon Angular Velocity

$$\Omega_H = \frac{ac}{2Mr_+}$$

- Superadiance Condition

$$0 < \omega < m\Omega_H$$

- Hauntum Tempertature

$$T_H = \frac{hc^6}{8\pi G M k_B}$$

Classical depends on environment.
Quantum depends only on M.

11. Key Takeaways

- Black holes can send energy out in three main ways: magnetic fields, matter, and radiation.
- The Blandford–Znajek mechanism taps rotational energy via magnetic fields.
- Jets carry kinetic energy that shapes galaxies and clusters.
- Quantum radiation across the spectrum reveals the physical processes at work.
- From AGN to micro black holes, the same physics powers the Universe.
- Together, these channels connect black holes to the evolution of the cosmos.

The black hole becomes a cosmic dynamo.
Its spin energy lights up the Universe.

12. In one Sentence

> *"Whether through classical fields and matter or quantum fluctuations of spacetime, black holes convert curvature into energy—and shape the Universe at every scale."*

"One engine. Two realms. Infinite consequences."

29

Quantum Fields in Curved Spacetime

The Stage for Hawking Radiation

Quantum field theory in the curved spacetime around a black hole reveals that the vacuum is not empty — particle pairs form, and one escapes as Hawking radiation.

1. Curved Spacetime and Quantum Fields

Matter curves spacetime. Quantum fields live on this curved background.

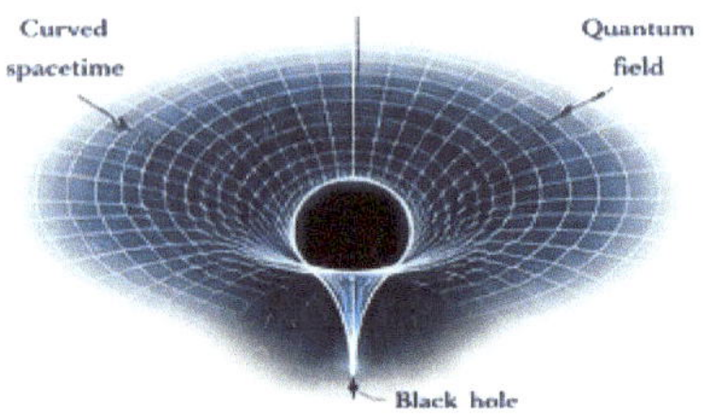

$$\Box_g \phi - m^2\phi = 0$$

Klein–Gordon equation in curved spacetime

The vacuum state depends on the observer. What is empty for one observer may not be empty for another.

2. Modes and Vacuum States

In curved spacetime, positive-frequency modes depend on time and position.

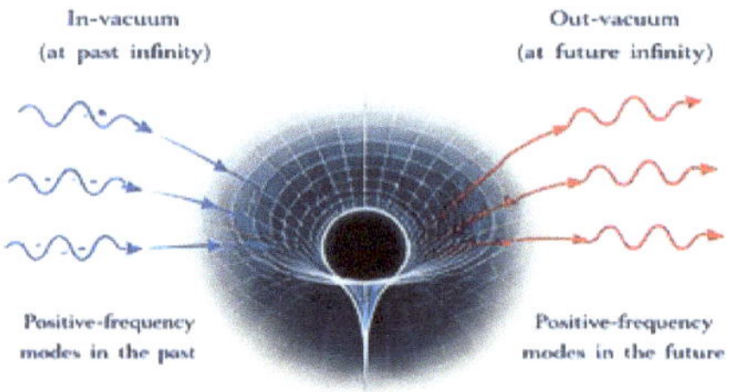

Gravitational collapse mixes positive and negative frequencies.

Modes are related by Bogoliubov transformations.

$$u^{out}_{\omega} = \sum_{\omega'} \left(\alpha_{\omega\omega'} u^{in}_{\omega'} + \beta_{\omega\omega'} u^{in*}_{\omega'}\right)$$

3. Pair Creation Near the Horizon

Vacuum fluctuations near the horizon can create particle-antiparticle pairs.

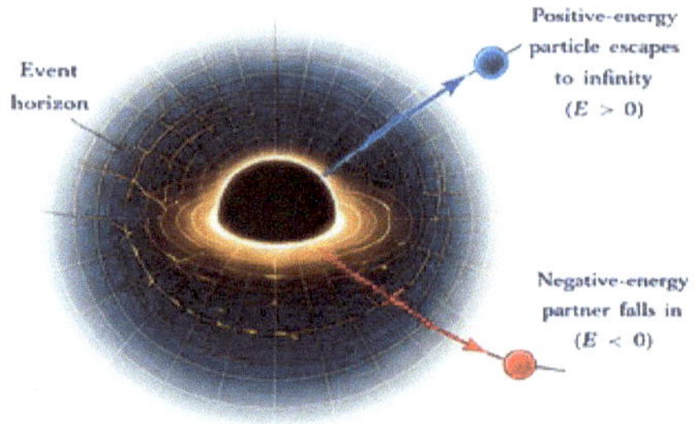

Energy is conserved: the black hole loses mass, and the escaping particle is seen as radiation.

4. The Essential Mechanism

1. Zero-point fluctuations exist everywhere.
2. Near the horizon, a pair is created.
3. One particle has negative energy and falls into the black hole.
4. The other has positive energy and escapes to infinity.

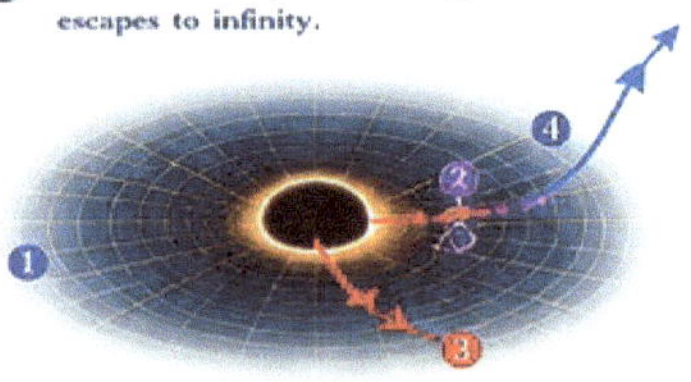

To a distant observer, this appears as thermal radiation from the black hole.

5. Hawking Temperature

The spectrum of particles at infinity is thermal with temperature

$$T_H = \frac{\hbar c^3}{8\pi G M k_B}$$

where

- M = black hole mass
- G = gravitational constant
- c = speed of light
- k_B = Boltzmann constant

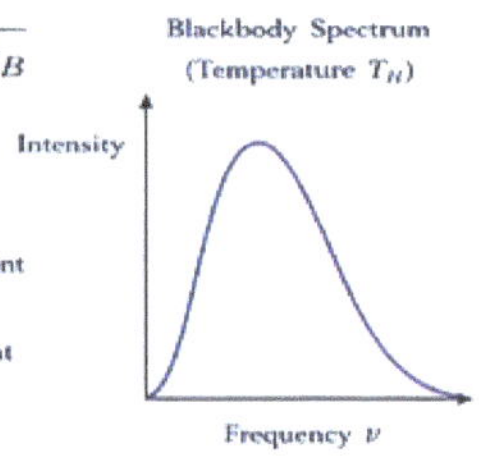

Larger black holes are colder.

Stellar-mass BH: $\sim 10^{-8}$ K   Supermassive BH: $\sim 10^{-10}$ K

6. Energy and Mass Loss

Hawking radiation carries energy away, so the black hole loses mass.

$$\frac{dM}{dt} = -\frac{\hbar c^4}{15360\pi G^2 M^2}$$

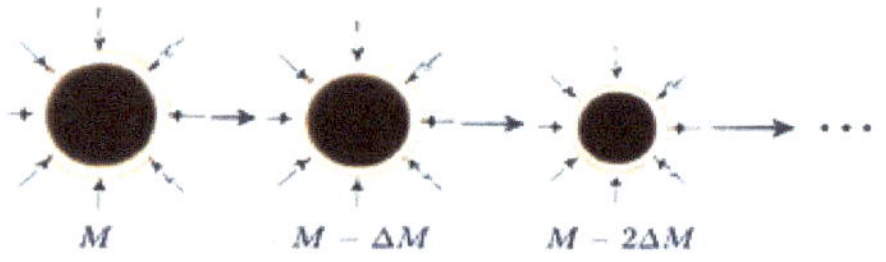

Over time, the black hole evaporates.

Smaller mass ⟶ higher temperature ⟶ faster evaporation.

7. In–Vacuum vs Out–Vacuum

The in–vacuum (no particles early on) is not the same as the out–vacuum (no particles late).

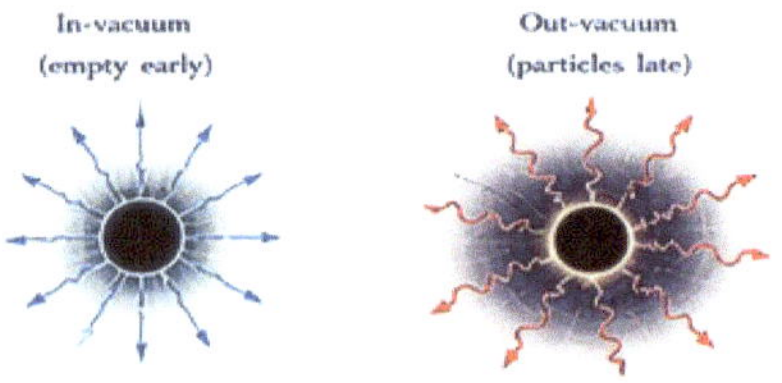

The initial vacuum evolves into a state with real particles at future infinity.

8. Analogy: Surfing the Horizon

Imagine waves on water flowing over a waterfall. Pairs form at the edge—one carried away, one falls in.

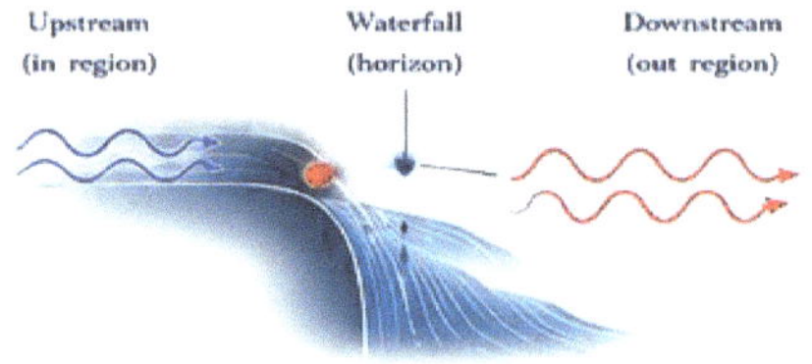

The flow creates pairs; the downstream wave escapes, the upstream one disappears.

9. Key Observational Implications

- **Relativistic jets** — Detected in radio, X-ray, and gamma-ray.
- **Quasi-Periodic Oscillations** — Reveal dynamics near the horizon.
- **Polarized emission** — Traces magnetic fields and jet launching.
- **Thermal & Non-thermal Spectra** — Show multiple energy components.
- **Evaporation over time** — Possible in γ-ray background?

Direct detection is extremely difficult, but the effects are fundamental.

10. Key Equations Summary

- Klein–Gordon equation:

$$\Box_g \phi - m^3\phi = 0$$

- Bogoliubov transformation:

$$u^{out}_{\omega} = \sum_{\omega} \left(\alpha_{\omega\omega'} u^{in}_{\omega'} + \beta_{\omega\omega'} u^{in*}_{\omega'}\right)$$

- Hawking temperature:

$$T_H = \frac{\hbar c^3}{8\pi G M k_B}$$

- Mass loss rate:

$$\frac{dM}{dt} = -\frac{\hbar c^4}{15360\pi G^2 M^2}$$

- Evaporation time:

$$\tau \sim \frac{5120\pi G^2 M^3}{\hbar c^4}$$

Different formulas, same engine—the quantum physics of curved spacetime.

11. Key Takeaways

- Spacetime curvature makes the vacuum active.
- Positive and negative frequency modes mix.
- Vacuum fluctuations near the horizon create pairs.
- One falls in with negative energy; the other escapes with positive energy.
- The radiation is thermal with temperature T_H.
- Black holes lose mass and eventually evaporate.
- This connects gravity, quantum theory, and thermodynamics.

Hawking radiation shows that black holes are not completely black—they shine.

12. In One Sentence

> *“Quantum fields in curved spacetime turn the vacuum into a source of radiation, letting black holes slowly evaporate through the creation of particle pairs.”*

“The vacuum is not empty near a black hole—it is alive with possibility.”

30 QUANTUM ENERGY BALANCE

HOW BLACK HOLES EXTRACT, RADIATE, AND CONSERVE ENERGY

Black holes do not break the laws of physics—quantum processes near the horizon ensure a delicate balance of energy in, out, and stored.

1. The Big Picture

Energy flows around a black hole in many ways. The total balance is always conserved.

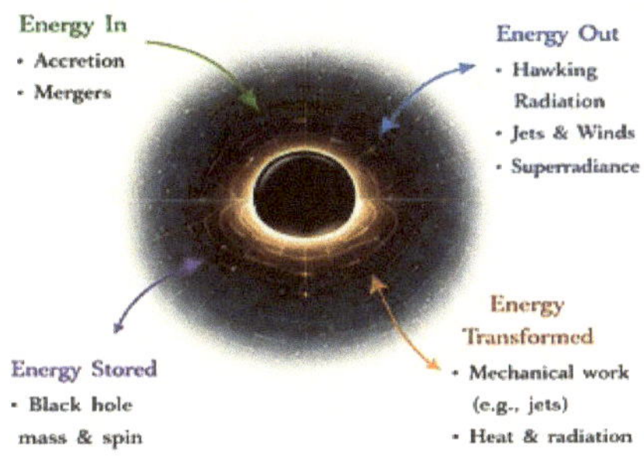

All forms of energy are accounted for. The total never disappears.

2. Energy Components

Mass–energy resides in the black hole

$$E_{BH} = Mc^2$$

Rotational energy (if spinning)

$$E_{rot} = \left(1 - \frac{\sqrt{1-a^2}}{2}\right) Mc^2$$

Extractable (maximal) rotational energy

$$E_{rot,max} \approx 0.29\, Mc^2 \quad (a \to 1)$$

Hawking radiation (quantum emission)

$$P_H = \frac{\hbar c^6}{15360 \pi G^2 M^2}$$

Accretion power (approx.)

$$P_{acc} = \eta \dot{M} c^2 \quad (0 < \eta < 0.42)$$

Energy comes in, is stored, converted, or radiated away.

3. Energy Flow Diagram

A schematic of quantum energy balance.

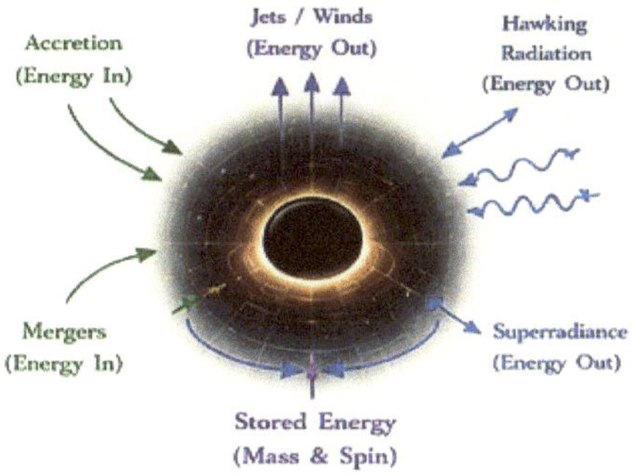

Quantum fields near the horizon redirect and convert energy in remarkable ways.

4. Quantum Origin: Pair Creation

Vacuum fluctuations create particle–antiparticle pairs. One falls in, the other escapes.

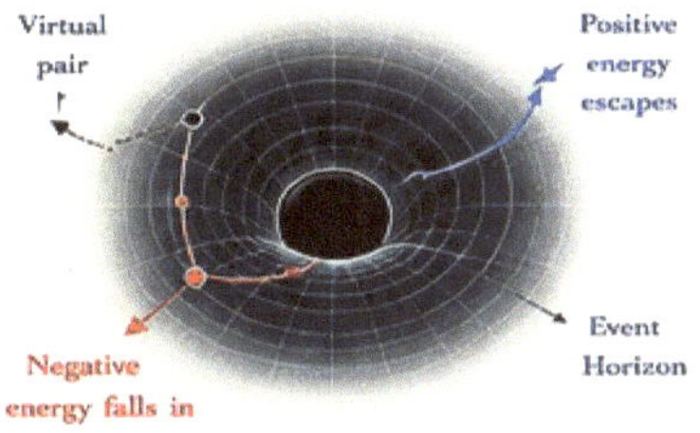

The escaping particle becomes Hawking radiation. The black hole loses mass.

5. Energy Conservation Equation

The total energy balance must satisfy:

$$\frac{dE_{tot}}{dt} = P_{in} - P_{out} - P_{work} = 0$$

Where:

- P_{in} = energy in (accretion, mergers)
- P_{out} = energy out (Hawking radiation, superradiance, winds, jets)
- P_{work} = useful work done on surroundings

Energy is neither created nor destroyed— it is transformed.

6. The Black Hole Lifetime

Smaller black holes evaporate faster.

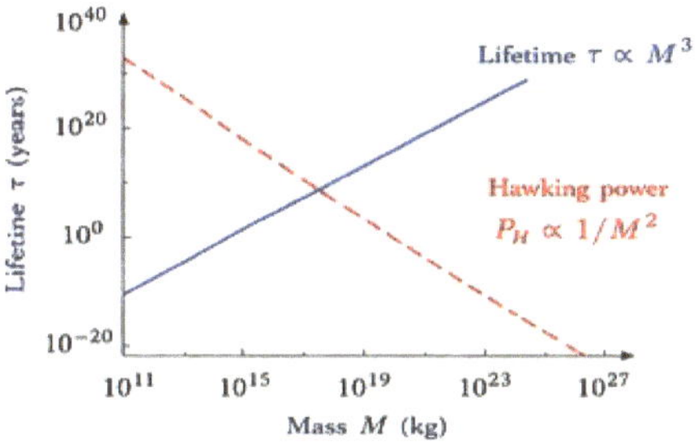

Stellar-mass BHs live longer than the Universe. Tiny black holes evaporate in an instant.

7. Rotating Black Holes: Extraction

Rotational energy can be extracted by multiple mechanisms.

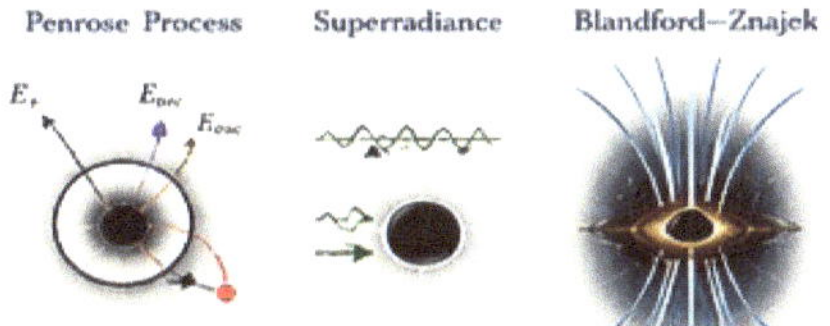

Energy comes from the black hole's spin, not just from accretion.

8. Typical Energy Budget (Active AGN)

Fractional power output (order of magnitude).

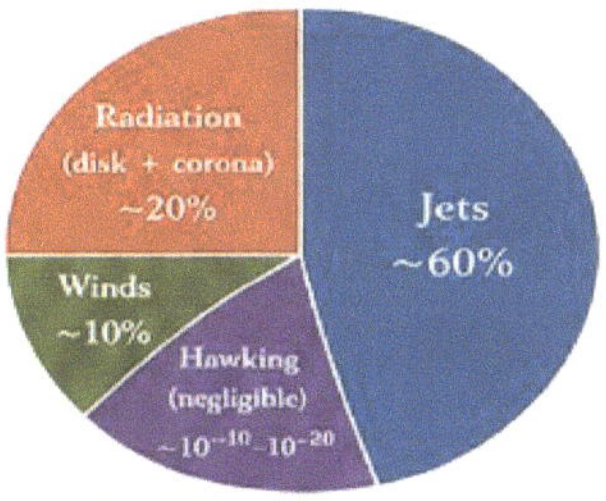

In most astrophysical black holes, Hawking radiation is tiny but fundamental.

9. Scales of Quantum Effects

Quantum effects are strongest for small masses and high curvatures.

Mass (kg)	Hawking Temp. T_H (K)	Hawking Power P_H (W)	Lifetime
10^{12} (Lunar)	10^{11}	10^{17}	10^{-27} s
10^{15} (Asteroid)	10^{8}	10^{11}	10^{-18} s
10^{20} (Stellar BH)	10^{-5}	10^{-28}	$> 10^{67}$ yr
10^{30} (Supermassive)	10^{-14}	10^{-52}	$> 10^{100}$ yr

Quantum effects are negligible for large black holes, but crucial for the smallest ones.

10. Energy Transformations

Energy changes form, but stays in the books.

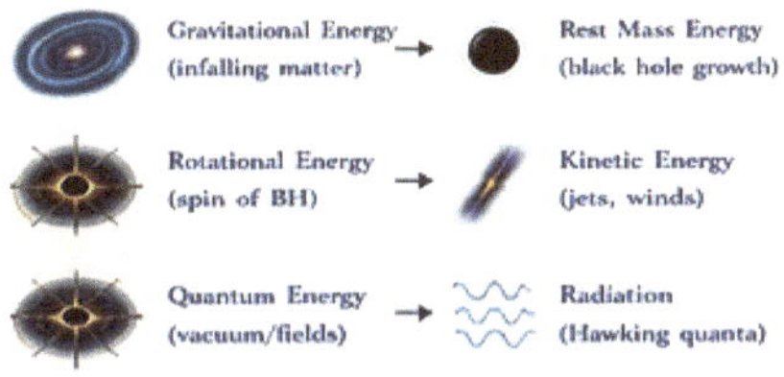

A cosmic accountant keeps perfect records. General relativity + quantum theory agree.

11. Key Takeaways

- Black holes obey energy conservation.
- Energy comes in, is stored, transformed, or radiated out.
- Quantum fields near the horizon create Hawking radiation.
- Spin energy can be extracted efficiently.
- Hawking radiation is tiny for large BHs, but central to theory.
- Every black hole has an energy budget.
- The Universe's ledger always balances.

Black holes are not energy sinks— they are dynamic engines of the cosmos.

12. In one Sentence

> *"Black holes never violate the rules— they convert, radiate, and balance energy through the deepest quantum processes in nature."*

From quantum fluctuations to cosmic jets—energy flows, transforms, and balances. The Universe keeps perfect accounts.

31

BEYOND HAWKING RADIATION

OTHER QUANTUM AND NON-THERMAL EFFECTS NEAR BLACK HOLES

Hawking radiation is only one expression of quantum physics near black holes.
Many other mechanisms can extract or release energy.

1. BEYOND THERMAL: A WIDER PICTURE

Black holes support many processes beyond thermal radiation.

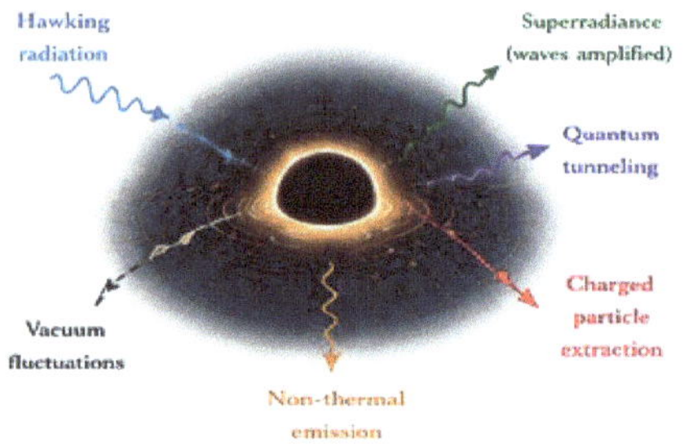

Different quantum effects, same source: curved spacetime and horizons.

2. SUPERADIANCE (REVISITED)

Waves with $0 < \omega < m\Omega_H$ get amplified when scattered by a rotating black hole.

Condition:
$0 < \omega < m\Omega_H$

Energy and angular momentum are extracted from the black hole's rotation.

3. QUANTUM TUNNELING

Particles can tunnel through the potential barrier near the horizon.

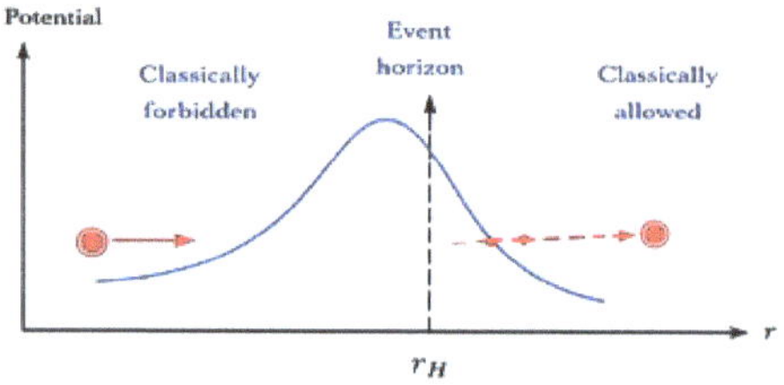

Tunneling probability:

$$P \sim e^{-2\int_{r_1}^{r_2} \kappa(r)dr}$$

Leads to particle emission and pair creation.

4. VACUUM FLUCTUATIONS AND PAIRS

The quantum vacuum is never empty.
Fluctuations create virtual pairs that can separate.

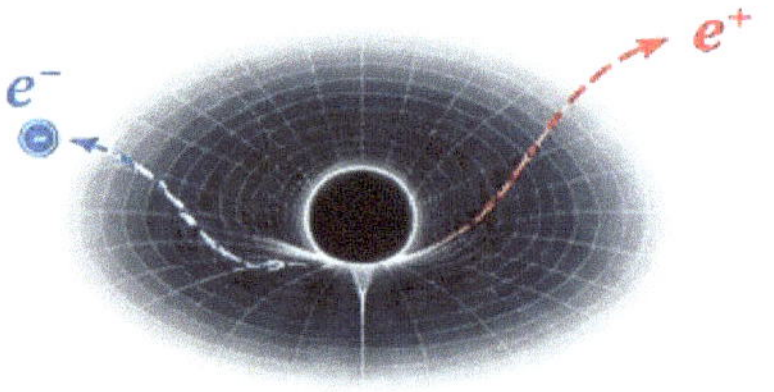

One falls in (negative energy), the other escapes (positive energy).

This reduces the black hole's mass.
Source of Hawking radiation.

5. CHARGED PARTICLE EXTRACTION

In electromagnetic fields, charged particles can gain energy and escape.

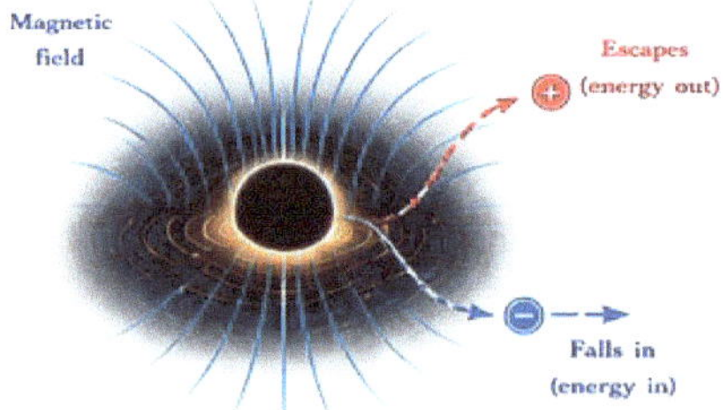

Similar in spirit to the Penrose process, but driven by electromagentic fields.

6. NON-THERMAL EMISSION

Accretion disks, magnetic reconnection, and shocks produce high-energy emission.

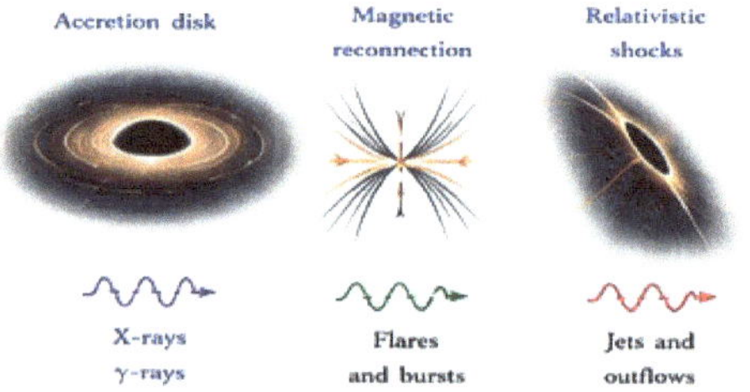

Dominant in astrophysical black holes.
Powers AGN, quasars, and X-ray binaries.

7. QUASI-NORMAL MODES (QNMs)

After a disturbance, the black hole "rings" with characteristic damped oscillations.

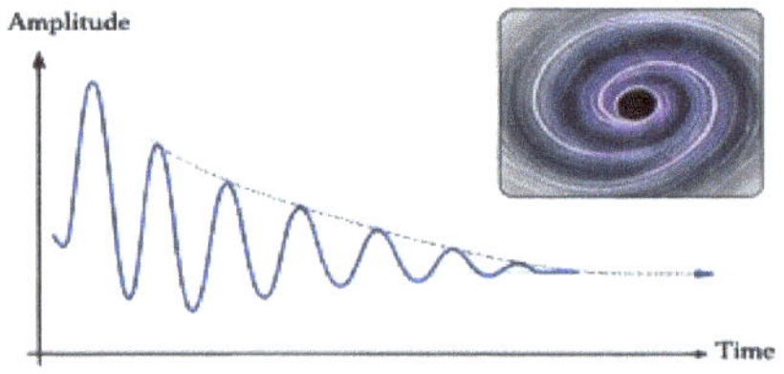

$f_{QNM} = f(M, a, charge, field)$

QNMs carry information about the black hole.
Detected in gravitational waves.

8. BLACK HOLE LASERS

Superradiant amplification between a black hole and a reflecting boundary creates a "laser".

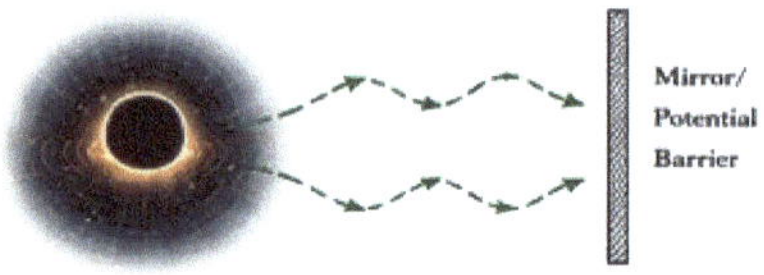

Instability condition:
$0 < \omega < m\Omega_H$ and sufficient reflectivity

A macroscopic quantum phenomenon with possible astrophysical relevance.

9. MEMORY AND INFORMATION?

Quantum effects near black holes raise deep questions about information and unitarity.

Possibilities

- Information escapes in subtle correlations (e.g., radiation).
- Information is stored on the horizon (firewall, etc.).
- Information emerges through new physics.

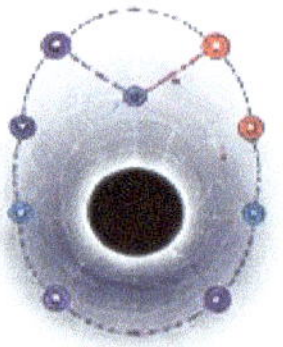

Open Questions

- How does information avoid being lost?
- Is spacetime emergent from quantum entanglement?
- What is the ultimate description?

The full theory must unite quantum mechanics, gravity, and information.

10. KEY FORMULAS

Hawking temperature	$T_H = \frac{\hbar c^3}{8\pi G M k_B}$
Superradiance condition	$0 < \omega < m\Omega_H$
Hawking power (approx.)	$P_H = \frac{\hbar c^6}{15360\pi G^2 M^2}$
Tunneling probability	$P \sim e^{-2\int_{r_1}^{r_2} \kappa(r)dr}$
Evaporation time	$\tau \sim \frac{5120\pi G^2 M^3}{\hbar c^4}$

Different expressions, same message:
black holes radiate and interact quantumly.

11. KEY TAKEAWAYS

- Hawking radiation is thermal and universal.
- Superradiance extracts rotational energy.
- Tunneling and vacuum fluctuations create pairs.
- Magnetic fields enable charged particle extraction.
- Accretion and shocks produce non-thermal emission.
- Black holes ring in quasi-normal modes.
- Black hole lasers can cause instabilities.
- Information may be preserved in subtle ways.
- Together, these reveal the quantum life of black holes.

One object. Many faces.
Infinite physics.

12. IN ONE SENTENCE

"Black holes are not black in the quantum view—they shine, they amplify, they radiate, they evolve, and they challenge our deepest understanding of reality."

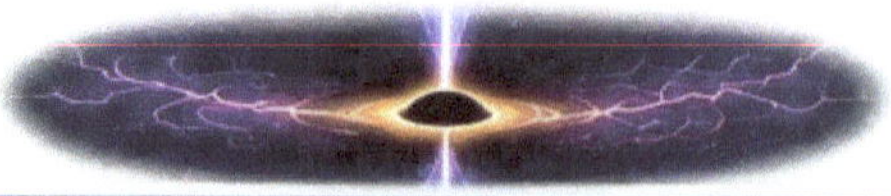

"Beyond the horizon lies not silence, but quantum activity—energy flows, transforms, and balances. The Universe keeps perfect accounts."

32 THE QUANTUM BRIDGE

FROM HAWKING RADIATION TO INFORMATION PARADOX

Hawking radiation reveals a deep mystery: information seems to disappear.
This section builds the bridge from radiation to the paradox.

1. The Setting

A black hole radiates thermally, as if it were a black body.

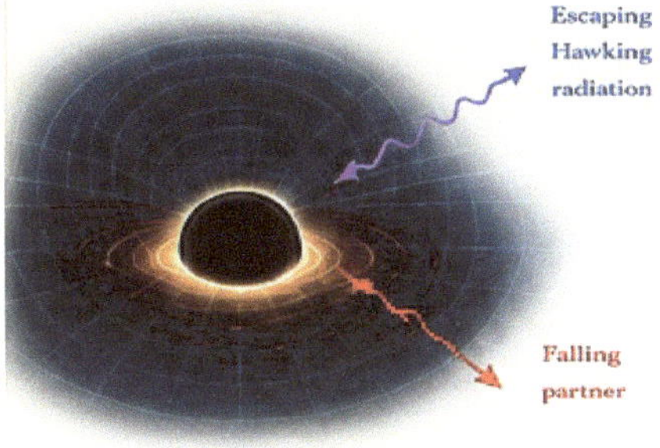

Near the horizon, quantum pairs form. One escapes; the other falls in. This process repeats constantly.

2. Entangled Pairs

The Hawking pair is born in an entangled state. They share quantum information.

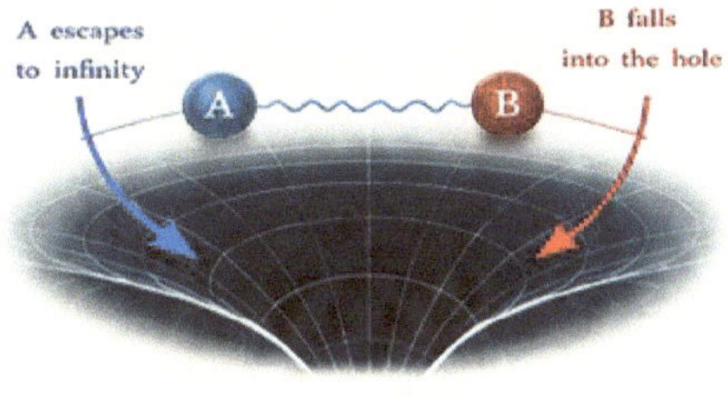

State: $|\Psi_{AB}\rangle = \sum_i c_i \, |i\rangle_A \, |i\rangle_B$

Tracing over B (inside) leaves A in a mixed state—thermal radiation.

3. Accumulation of Radiation

Over time, many pairs are created. Radiation builds up outside.

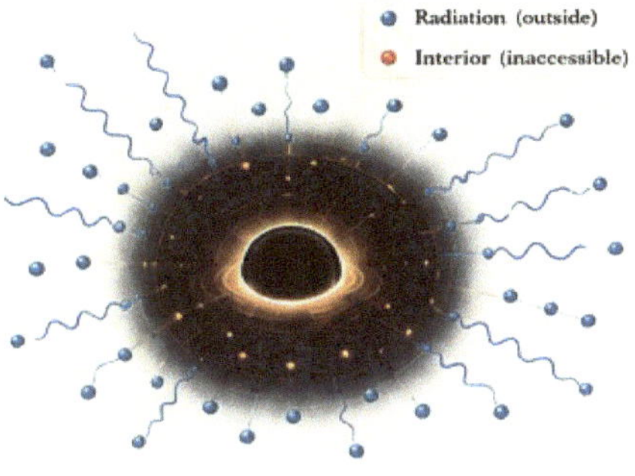

The outside observer sees a growing bath of seemingly thermal radiation.

4. The Information Question

If radiation is thermal, it carries no information about what fell in.

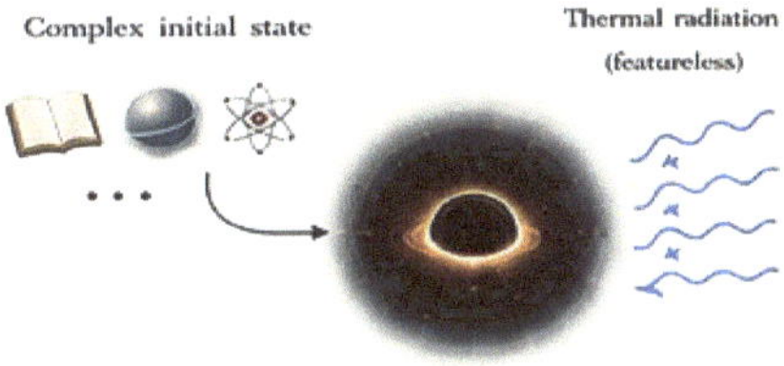

Information about the initial state appears to be lost forever beyond the horizon.

5. Unitarity vs. Thermality

Quantum mechanics demands unitarity. But Hawking radiation looks thermal.

Quantum Mechanics (Unitary)		Hawking Picture (Thermal)
Information is preserved.	VS.	Information is lost.
$|\psi_{in}\rangle \rightarrow |\psi_{out}\rangle$ (pure state)		$\rho_{out} = \sum_{i,j} p_i \, |\psi_j^{(i)}\rangle\langle\psi_j^{(i)}|$ (mixed state)

These two principles seem to clash. This is the heart of the paradox.

6. Page Curve Preview

If evolution is unitary, the entropy of radiation should eventually decrease.

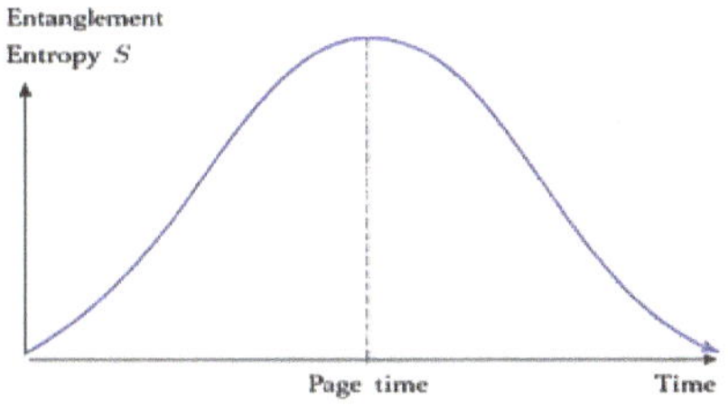

At the Page time, half the black hole has evaporated. After that, S should fall.

7. Where the Paradox Begins

Three assumptions lead to the paradox.

1. Hawking radiation is exactly thermal.
2. Nothing escapes from behind the horizon.
3. Quantum mechanics is unitary.

These cannot all be true—at least one must give way.

8. Possible Resolutions (Preview)

Many bold ideas have been proposed. We will explore them later.

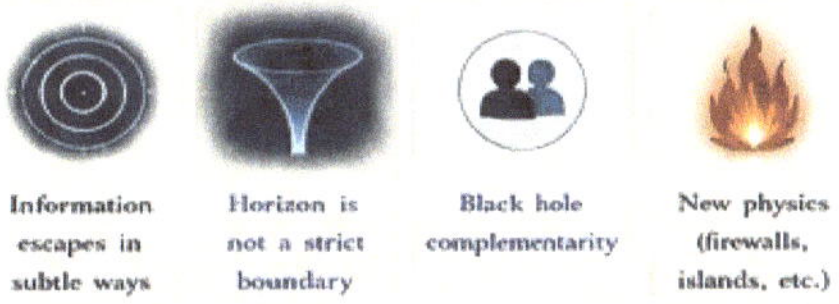

Different ideas—same goal: save both quantum mechanics and gravity.

9. The Road Ahead

The paradox is not a problem to ignore, but a signpost to deeper truth.

- We will examine information flow.
- We will study holography and entanglement.
- We will seek a complete theory of quantum gravity.

The bridge from radiation to paradox leads us toward the theory of everything.

10. Key Equations

Hawking temperature	$T_H = \dfrac{\hbar c^3}{8\pi G M k_B}$
Superradiance condition	$0 < \omega < m\Omega_H$
Hawking power (approx.)	$P_H \simeq \dfrac{\hbar c^6}{15360\pi G^2 M^2}$
Tunneling probability	$P \sim e^{-2\int_{r_1}^{r_2} \kappa(r')\,dr'}$
Evaporation time	$\tau \sim \dfrac{5120\pi G^2 M^3}{\hbar c^4}$

These are the tools that reveal the depth of the paradox.

11. Key Takeaways

- Hawking radiation arises from quantum effects.
- It appears thermal and carries no obvious information.
- Quantum entanglement links radiation and interior.
- Information seems lost—but unitarity says it cannot be.
- The Page curve suggests information must return.
- The paradox challenges our understanding of reality.
- Solving it will require new ideas about space, time, and quantum gravity.

One object. Many faces. Infinite physics.

12. In One Sentence

> *“Hawking radiation opened the door; the information paradox stands in the doorway, guiding us to the deeper theory beyond.”*

The paradox is the gateway. The answer lies beyond—to understand the Universe, we must understand itself.

33

THE INFORMATION PARADOX

WHY INFORMATION SHOULD NOT BE LOST

Hawking radiation appears thermal, but quantum mechanics demands information is preserved.
This tension is the information paradox.

1. The Puzzle

Hawking radiation is exactly thermal—
it carries no information about what fell in.

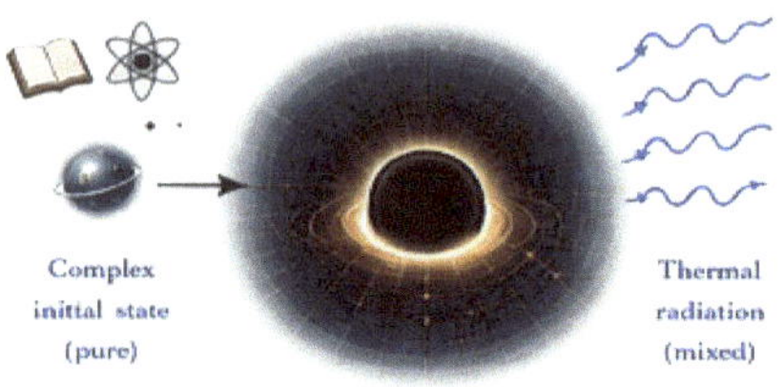

Pure state ⟶ Mixed state ?

This seems to violate unitarity.
Where does the information go?

2. What Unitarity Demands

Quantum evolution must be unitary.
Pure states must remain pure.

Evolution:

$$|\psi_{in}\rangle \xrightarrow{U} |\psi_{out}\rangle$$

where U is unitary: $U^{\dagger}U = I$

von Neumann entropy

$$S = -\mathrm{Tr}(\rho \ln \rho)$$

Pure state: $\rho = |\psi\rangle\langle\psi|$, $S = 0$ → Unitary evolution U → Pure state: $\rho' = U\rho U^{\dagger}$, $S' = 0$

Information cannot be destroyed—
it can only be hidden.

3. Where Could Information Hide?

Information might be encoded in subtle
quantum correlations.

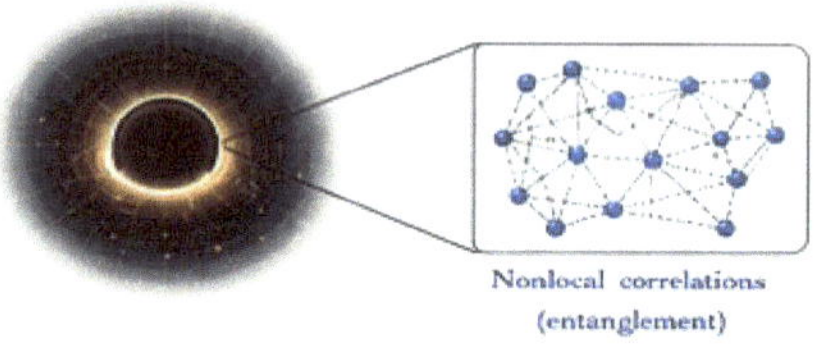

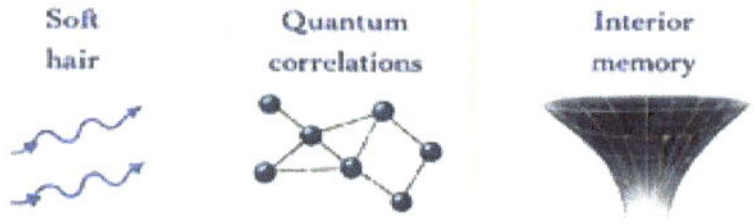

Information may be spread across many
modes—too subtle to detect.

4. Entanglement and Radiation

Hawking pairs are entangled.
Tracing out partners gives thermal radiation.

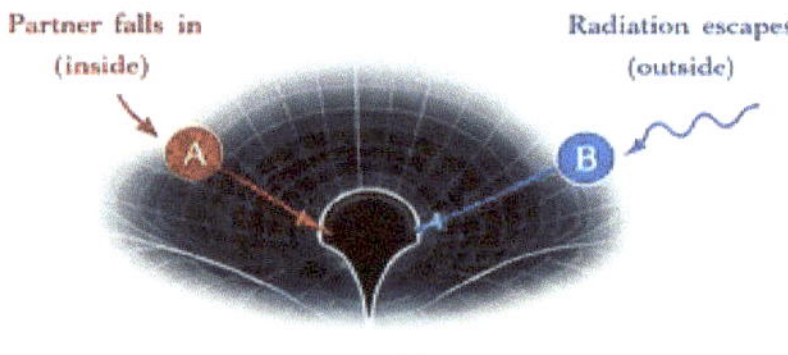

Entangled pair state:

$$|\Psi\rangle = \frac{1}{\sqrt{2}}\left(|0\rangle_A\,|1\rangle_{B\,\text{out}} + |1\rangle_A\,|0\rangle_{B\,\text{out}}\right)$$

Each pair alone looks thermal,
but together they are pure.

5. The Growing Entanglement

As radiation escapes, entanglement entropy
increases—until the Page time.

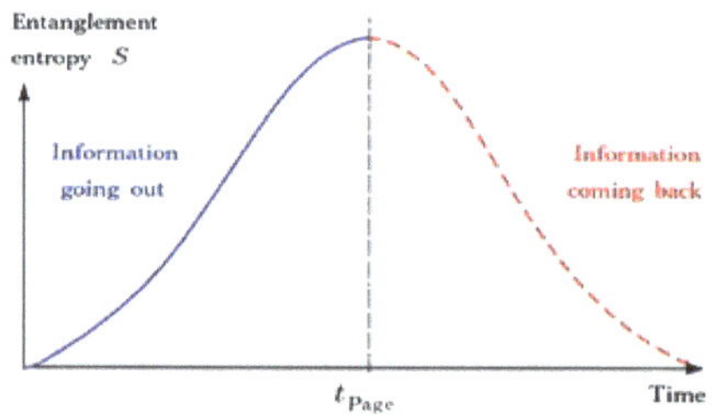

Before t_{Page}: information is lost to the outside.
After t_{Page}: it must return.

6. The Paradox

If radiation is truly thermal forever,
entropy never decreases.
This contradicts unitarity.

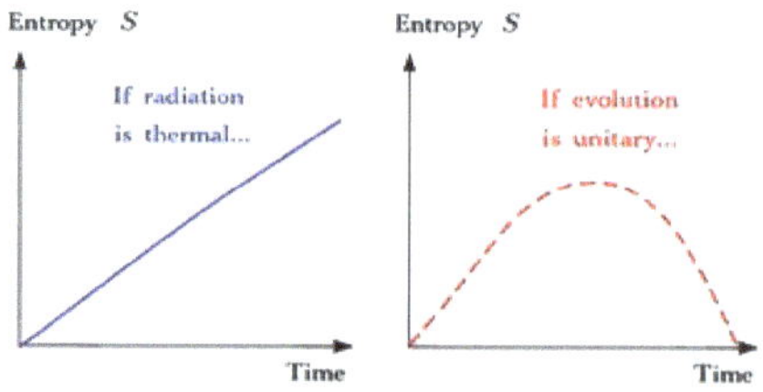

Which is correct?
Thermal forever—or unitary evolution?
The information paradox.

7. Proposed Ideas

Many bold ideas attempt to resolve the paradox.

- Information in the radiation (quantum correlations) — Information is encoded in highly entangled patterns.
- Black hole complementarity (no paradox) — No observer sees a violation of the laws of physics.
- Firewalls (drastic interior change) — Information emerges, but at the cost of a firewall.
- Remnants (Planck-scale relics) — Information stored in tiny, long-lived remnants.

No idea is proven—
the paradox remains open.

8. Holographic Insight

The holographic principle suggests that information
is stored on a lower-dimensional boundary.

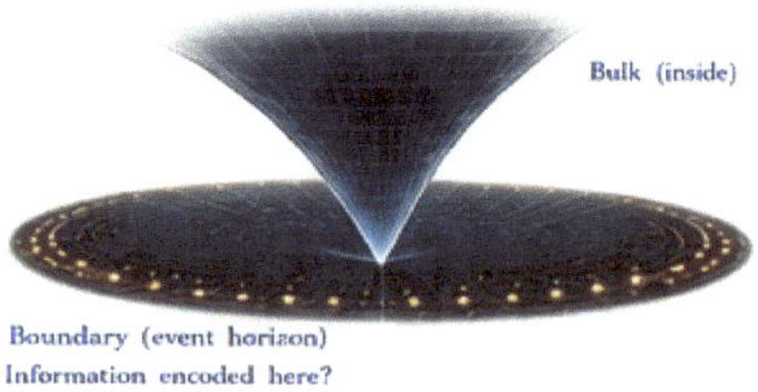

The interior may be encoded on the horizon—
like a hologram.

9. Islands and New Perspectives

Recent ideas show that including "islands" inside
the black hole can restore unitarity.

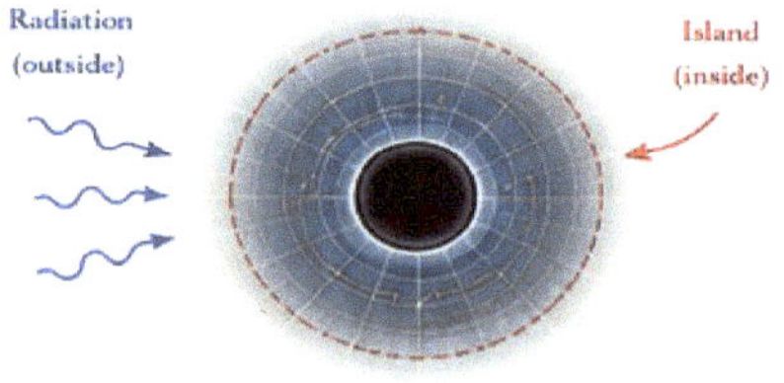

Quantum gravity may connect the inside and
outside in surprising ways.

10. Key Equations

- Hawking temperature $T_H = \dfrac{\hbar c^3}{8\pi G M k_B}$
- Hawking power (approx.): $P_H \approx \dfrac{\hbar c^6}{15360\pi G^2 M^2}$
- Entanglement entropy (Page curve peak): $S_{max} \approx \dfrac{A}{4\ell_P^2}$ (Bekenstein–Hawking)
- Unitary evolution: $|\psi_{in}\rangle \xrightarrow{U} |\psi_{out}\rangle, \quad U^{\dagger}U = I$

Thermodynamics describes the flow.
Quantum mechanics demands the truth.

11. Key Takeaways

- Hawking radiation arises from quantum effects.
- It appears thermal and carries no obvious information.
- Quantum entanglement links radiation and interior.
- Information seems lost—but unitarity says it cannot be.
- The Page curve suggests information must return.
- The paradox challenges our understanding of reality.
- Solving it will require new ideas about space, time, and quantum gravity.

One object. Many faces.
Infinite physics.

12. In one Sentence

> *"The information paradox is the bridge between quantum mechanics and gravity—and the key to a deeper theory of reality."*

Information is not lost—just hidden in the quantum deep.
The paradox is the gateway. The answer lies beyond.

34

THE PAGE CURVE

ENTANGLEMENT ENTROPY AND THE FATE OF INFORMATION

The Page curve describes how entanglement entropy of Hawking radiation rises, peaks, and then falls—if evolution is unitary.

1. THE QUESTION

Does a black hole destroy information, or is it returned in the radiation?
The Page curve gives a clue.

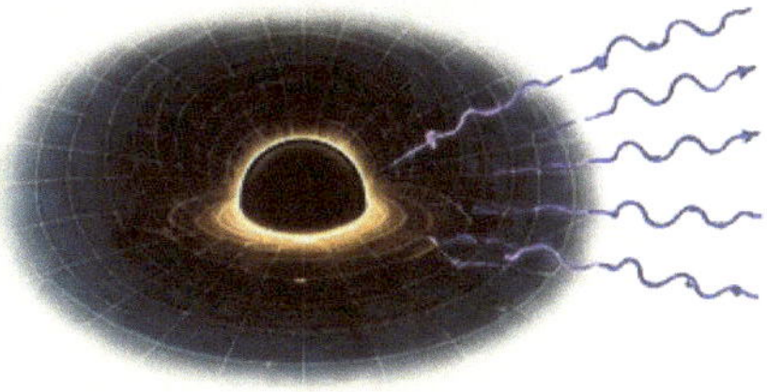

If information is lost, entropy keeps rising.
If information is preserved, entropy must fall.

2. ENTANGLEMENT ENTROPY

Consider the total system:
Black hole (B) + Radiation (R) = Closed system.
The entanglement entropy of radiation S_R measures how much it is entangled with B.

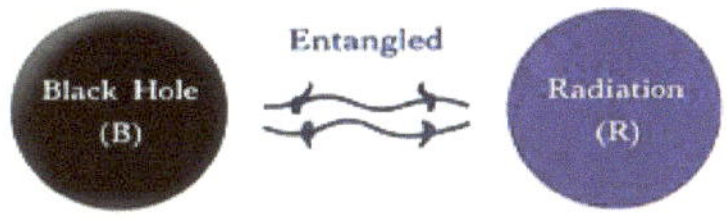

$$S_R = -\mathrm{Tr}(\rho_R \ln \rho_R)$$

where ρ_R is the reduced density matrix of the radiation.

3. TIME EVOLUTION OF EVAPORATION

As the black hole evaporates, radiation carries away energy and entropy.
The system evolves in time.

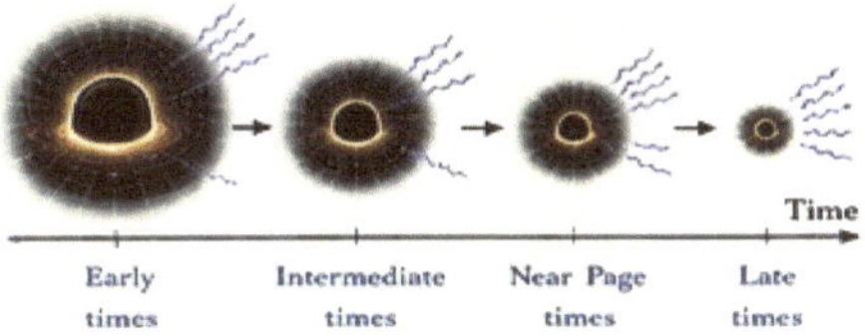

At first, radiation is entangled with the black hole. Later, radiation becomes entangled with earlier radiation.

4. THE PAGE CURVE

For a unitary evolution, the entropy follows a universal shape.

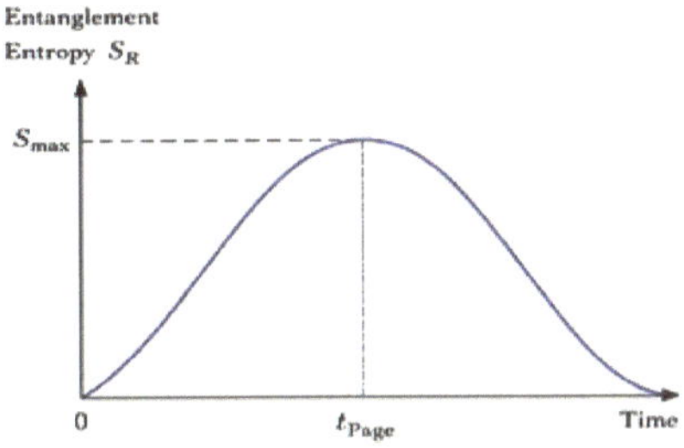

1. Early times Entropy rises
2. Around t_{Page} Entropy peaks
3. Late times Entropy falls

The Page time t_{Page} is when the black hole has lost about half of its initial entropy.
After that, information begins to come out.

5. PHYSICAL INTERPRETATION

1. Early times ($t < t_{Page}$)
Radiation is mostly entangled with the black hole.
Entropy increases.

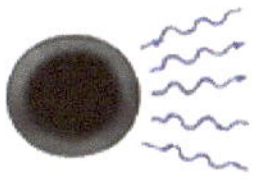

2. Page time ($t \approx t_{Page}$)
Radiation contains as much information about the early radiation as about the black hole.
Entropy is maximal.

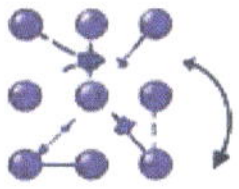

3. Late times ($t > t_{Page}$)
Radiation is entangled with earlier radiation, not the black hole.
Entropy decreases.

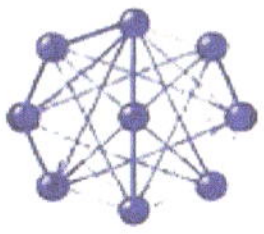

The radiation gradually purifies itself.
The final state is (almost) pure.

6. UNITARY VS. NON-UNITARY

Two possible outcomes:

Unitary evolution (information preserved)

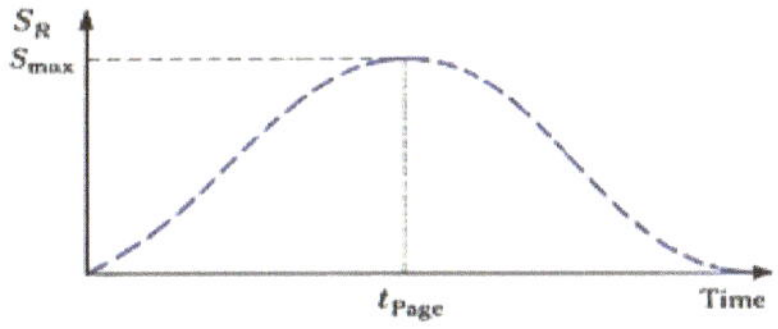

Non-unitary evolution (information lost)

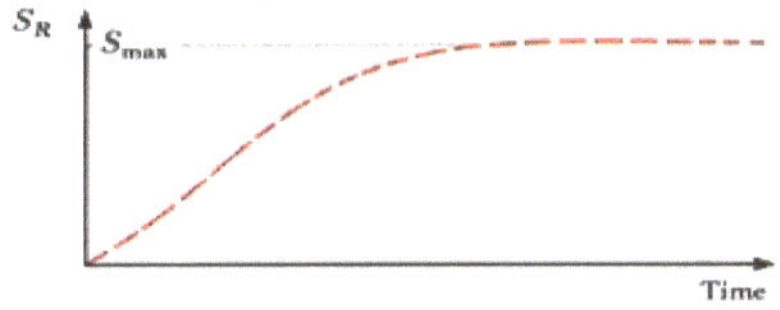

Only the first is consistent with quantum mechanics.

7. WHAT DETERMINES t_{Page} ?

Roughly, the Page time occurs when the radiation has entropy equal to half the initial black hole entropy.

$$S_R(t_{Page}) \approx \frac{1}{2} S_{BH,0}$$

For a black hole of initial mass M_0,

$$S_{BH,0} = \frac{4\pi G M_0^2}{\hbar c}$$

Thus $t_{Page} \sim \mathcal{O}(M_0^3)$
(in $G = c = \hbar = 1$ units)

8. INSIGHT

The Page curve links thermodynamics, quantum theory, and information.

Thermodynamics
Entropy counts disorder.

Quantum Theory
Unitarity preserves information.

Information Theory
Information is not destroyed.

Black holes may scramble information, but they do not erase it.

9. IMPLICATIONS

- Supports unitarity.
- Guides models of quantum gravity.
- Motivates ideas like black hole complementarity, holography, and quantum error correction.
- Constrains any theory that claims information is destroyed.

The Page curve is a beacon for resolving the information paradox.

10. MATHEMATICAL DETAILS (SKETCH)

For random unitary evolution (Page 1993), average entropy:

$$\langle S_R(t)\rangle \approx \begin{cases} t, & t < t_{Page} \\ S_{max} - (t - t_{Page}); & t > t_{Page} \end{cases}$$

where t is the entropy production rate.

After the Page time, entropy declines linearly (in a coarse-grained sense).

11. PAGE CURVE IN A NUTSHELL

Time Regime	Entropy Behavior	Meaning
Early times	Rises	Radiation learns about the hole
Around t_{Page}	Peaks	Maximal entanglement
Late times	Falls	Information is returned
Final state	Goes to zero	Radiation is (almost) pure

A beautiful curve that may hold the key to the fate of information.

12. IN ONE SENTENCE

> *"The Page curve tells us that if the Universe is unitary, black holes do not destroy information—they return it, hidden in the radiation."*

Information is not lost—just hidden in the quantum deep.
The paradox is the gateway. The answer lies beyond.

35

Modern Ideas on the Information Paradox

New Perspectives from Quantum Gravity and Beyond

The Page curve shows that information should return. Many recent ideas suggest how. Each offers a piece of the puzzle.

1. The Puzzle Remains

Hawking radiation looks thermal, but unitarity demands information is not lost.

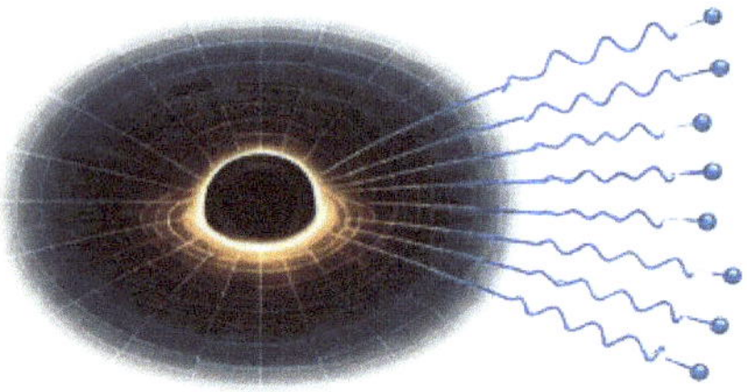

Thermal appearance vs. unitary reality — the core tension of the paradox.

2. The Page Curve Guide

The Page curve implies that information must begin to come out after the Page time t_{Page}.

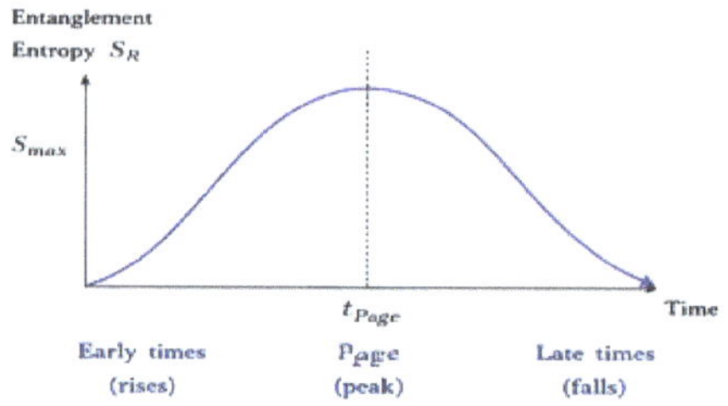

A unitary theory must reproduce this behavior.

3. Holographic Principle

Information is encoded on the boundary of spacetime, not in the bulk.

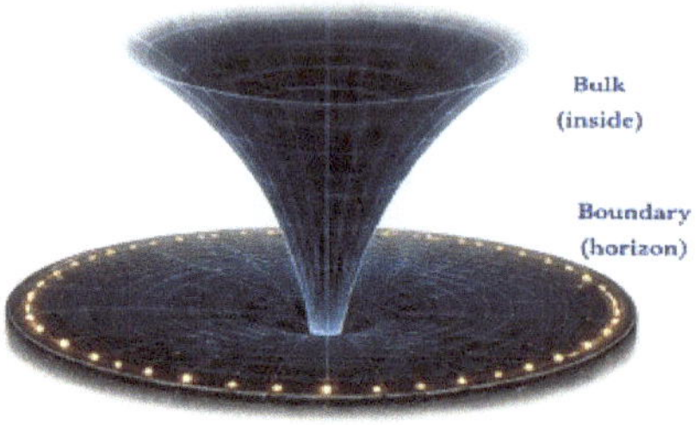

The boundary (horizon) stores the data — the bulk is a hologram.

4. Black Hole Complementarity

No single observer sees a violation. Different observers have different, but consistent, descriptions.

Sees smooth horizon and Hawking radiation.

Sees nothing unusual at the horizon.

Quantum mechanics is never violated, but reality is observer-relative.

5. ER = EPR Conjecture

Quantum entanglement (EPR) may be connected to spacetime geometry (ER).

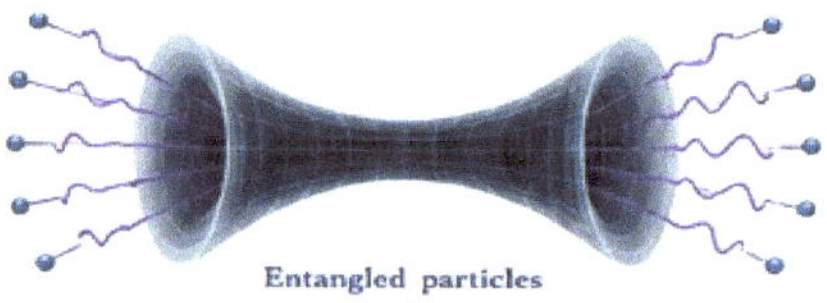

Entangled particles might be linked by a tiny wormhole.

Entanglement could be the glue that builds spacetime.

6. Quantum Error Correction (QEC)

Information is not lost — it is scrambled and protected in a highly redundant way.

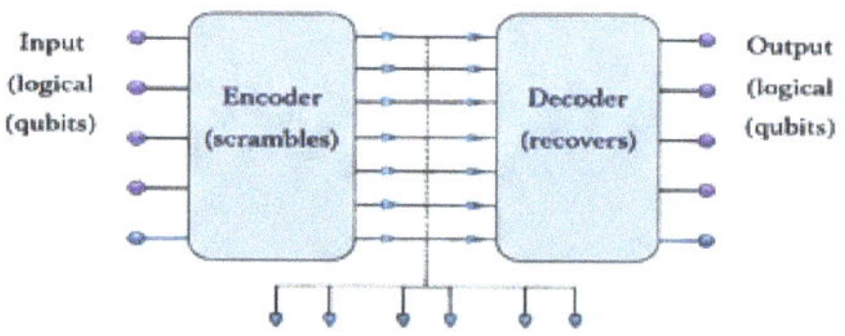

Bulk: degrees of freedom (many physical qubits)

The black hole acts like a quantum code: hard to read, but easy to recover.

7. Islands and Replica Wormholes

New calculations in quantum gravity show that "islands" inside the black hole can contribute to the entropy.

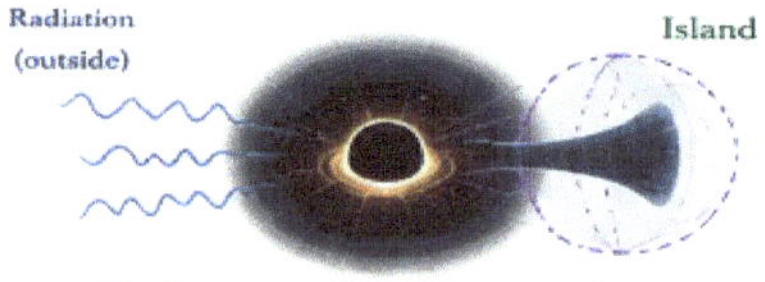

Replica wormholes connect the island to the radiation.

These effects can reproduce the Page curve in a unitary way.

8. Soft Hair & Memory

Blask holes might have subtle quantum structures ("soft hair") that carry information.

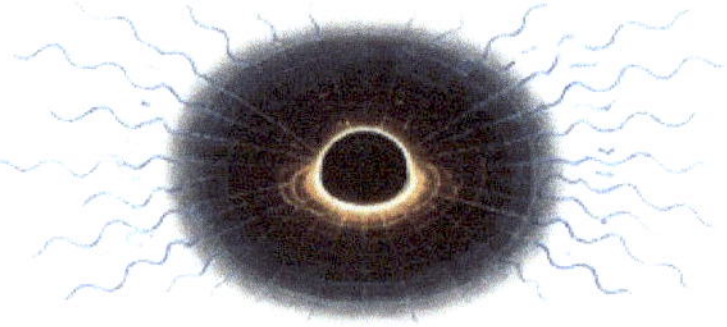

Soft modes on the horizon can store information.

Information could be hidden in very fine-grained horizon data.

9. Firewall Controversy

If information escapes, the equivalence principle may be in danger.

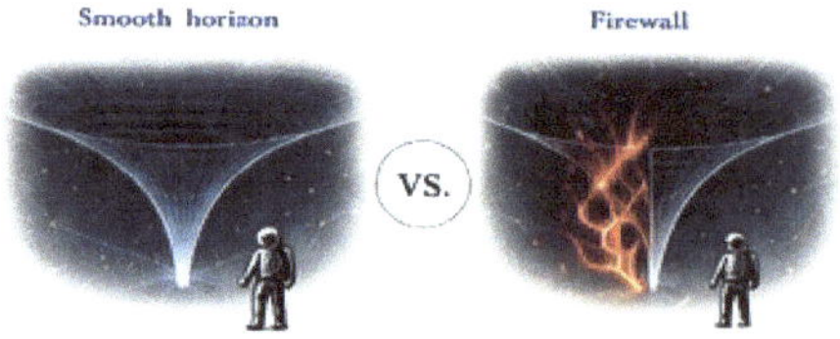

A radical possibility — still debated and not yet confirmed.

10. Key Ideas in Equations

- Entanglement entropy of radiation

$$S_R = -\mathrm{Tr}(\rho_R \ln \rho_R)$$

Measures how much radiation is entangled with the black hole.

- Page time (approx.)

$$t_{Page} \sim \frac{M^3}{G^2 \hbar c^4}$$

When half the black hole has evaporated.

- Quantum state evolution (unitarity)

$$|\psi_{in}\rangle \xrightarrow{U} |\psi_{out}\rangle, \quad U^\dagger U = I$$

Unitarity ensures information is preserved.

11. Key Takeaways

- The Page curve tells us information must return.
- Holography suggests information lives on the horizon.
- ER = EPR links entanglement to geometry.
- QEC shows how information can be hidden, not lost.
- Islands and replica wormholes explain the Page curve.
- Soft hair may store information in subtle ways.
- The paradox pushes us to a deeper theory of quantum gravity.

No single idea is final, but each brings us closer to the truth.

12. In One Sentence

> *The information paradox is not a dead end — it is a doorway to a deeper understanding of spacetime, quantum mechanics, and reality itself.*

The search for information is the search for reality.
The paradox is our map — the destination lies ahead.

36

Toward Quantum Gravity

How the Information Paradox Points the Way

The struggle to understand black hole information is not a dead end — it is the compass that may lead us to a deeper theory of nature.

1. The Paradox as a Guide

A paradox often signals that our current framework is incomplete.

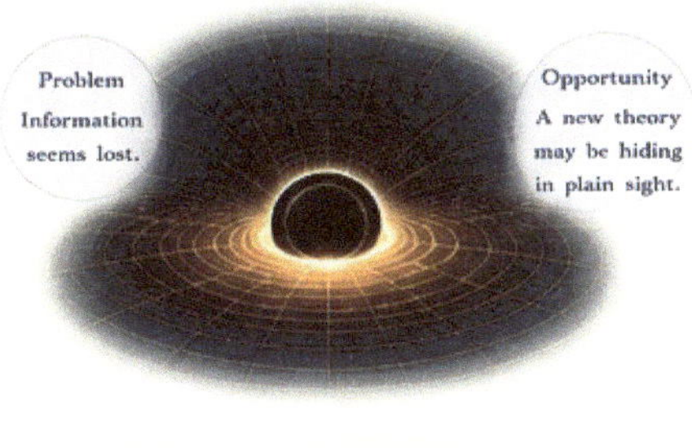

Nature is telling us:
Something deeper is waiting to be discovered.

2. What We Have Learned

The information paradox has already taught us profound lessons.

- Quantum effects are crucial even in strong gravity.
- Entanglement connects regions in ways we never imagined.
- Information is physical, not just abstract.
- Spacetime may emerge from quantum information.

These insights are shaping the path toward quantum gravity.

3. Candidate Resolutions

Several promising ideas aim to resolve the paradox.

Holographic Principle
Information is stored on the boundary. The bulk is a projection.

Islands & Replica Wormholes
Quantum connections create shortcuts that recover information.

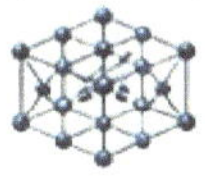

Quantum Error Correction (QEC)
The bulk encodes information redundantly, like a quantum code.

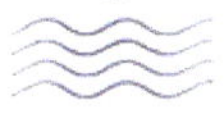

Soft Hair & New Symmetries
Hidden symmetries at the horizon may store information.

Different ideas, one goal: make information compatible with unitarity.

4. From Firewalls to Friendlier Horizons

The firewall proposal shocked us — but it pushed the community forward.

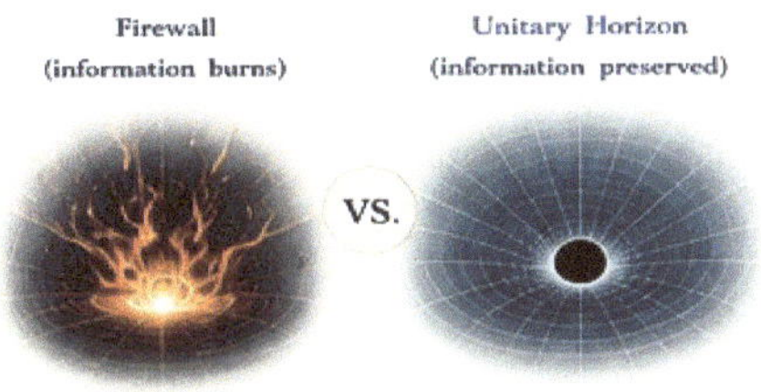

The search for a consistent, unitary horizon continues — and we are closer than before.

5. Beyond Semiclassical Thinking

Semiclassical gravity treats spacetime as fixed. The paradox shows this cannot be complete.

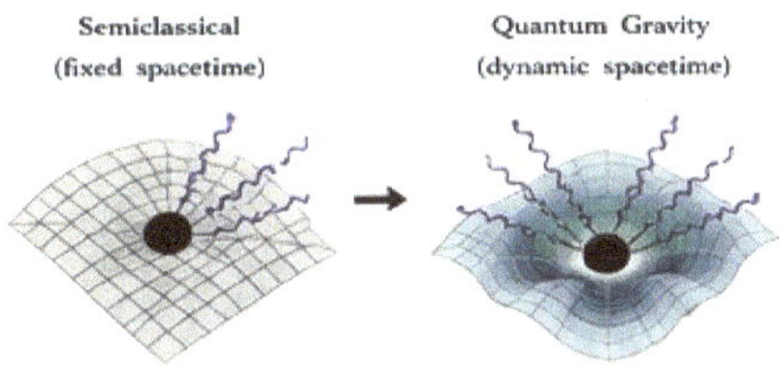

A full theory must include quantum spacetime and backreaction at all scales.

6. Building Blocks of the Future

Any successful theory should satisfy these principles.

1. Unitarity — information is never lost.
2. Equivalence Principle — gravity is universal.
3. Quantum Coherence — superposition survives.
4. Locality (or its deeper generalization).
5. Background Independence — spacetime is not fixed.

A theory that unites these will be our best description of reality.

7. New Mathematical Tools

The paradox has inspired powerful new mathematics.

- Topological Quantum Field Theories capture boundary/bulk relations.
- Tensor Networks model entanglement in quantum spacetime.
- Category Theory organizes holography and dualities.
- Non-perturbative techniques probe black hole microphysics.

Mathematics is becoming the language of quantum gravity.

8. Observational Windows

Future observations may reveal hints of information recovery.

Primordial black holes could show subtle imprints.

Gravitational waves probe strong gravity dynamics.

Black hole shadows & rings test spacetime structure.

High-energy cosmic rays may carry exotic signatures.

Observations can guide theory and theory can guide observations.

9. The Bigger Picture

The information paradox connects to the deepest questions.

What is spacetime?
Is it fundamental or emergent?

What is information?
Is it conserved, or created?

What is time?
Does it emerge from entanglement?

What is reality?
What is real at the most fundamental level?

The paradox is not just about black holes — it is about the nature of existence.

10. Key Ideas in Equations

- Entanglement entropy of radiation

$$S_R = -\mathrm{Tr}\,(\rho_R \ln \rho_R)$$

Measures how much radiation is entangled with the black hole.

- Page time (approx.)

$$t_{\text{Page}} \sim \frac{M^3}{G^2 \hbar c^4}$$

When half the black hole has evaporated.

- Quantum state evolution (unitary)

$$|\psi_{\text{in}}\rangle \xrightarrow{U} |\psi_{\text{out}}\rangle, \quad U^\dagger U = I$$

Unitarity ensures information is preserved.

11. Key Takeaways

- The Page curve tells us information must return.
- Holography suggests information lives on the horizon.
- ER = EPR links entanglement to geometry.
- QEC shows how information can be hidden, not lost.
- Islands and replica wormholes explain the Page curve.
- Soft hair may store information in subtle ways.
- The paradox pushes us to a deeper theory of quantum gravity.

No single idea is final, but each brings us closer to the truth.

12. In One Sentence

> *“The information paradox is not a dead end — it is a gateway to a deeper understanding of spacetime, quantum mechanics, and reality itself.”*

The search for information is the search for reality.
The Page curve is our map — the destination lies ahead.

37

Future Energy Extraction Technologies

From Concepts Today to Civilizations Tomorrow

If nature allows energy to be extracted, future civilizations may learn how—
safely, wisely, and in harmony with the laws of the Universe.

1. Why Look to the Future?

Black holes are extreme engines.
If we understand them,
we may one day harness them.

Understanding today becomes
technology tomorrow.

2. Guiding Principles

Any future technology must obey:

- Physical laws (causality, thermodynamics)
- Information preservation (unitarity)
- Sustainability (no catastrophic instability)
- Ethics & stewardship (for all life and future generations)

Power without wisdom
destroys. Power with wisdom elevates.

3. Advanced Magnetohydrodynamic Systems

Ultra-efficient magnetic structures may tap
rotational energy via the Blandford–Znajek process.

Stronger fields, smarter control,
and self-regulating plasmas.

4. Miniaturized Ergosphere Engineering

Artificial structures might interact with the
ergosphere to extract energy safely.

Dynamic tethers or field interactors
ride negative energy states.

5. Hawking Radiation Harvesters

Future materials could capture Hawking quanta
without disturbing the horizon.

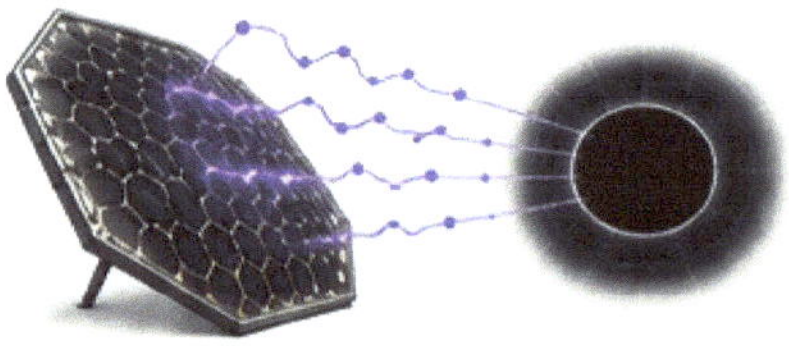

Quantum sensors, metamaterials,
and low-noise detectors.

6. Quantum Energy Converters

Energy extracted at the quantum level
may be converted with near-perfect efficiency.

From quantum fluctuations
to usable macroscopic power.

7. Space-Based Megastructures

Dyson-like or adaptive megastructures
could orbit black holes at safe distances.

Collect, convert, and transmit
energy across interstellar distances.

8. Information-Safe Extraction

Technologies must guarantee
information is never destroyed.

Built-in quantum error correction,
holographic safeguards, and redundancy.

9. Self-Regulating Systems

Future devices will adapt in real time
to preserve stability and safety.

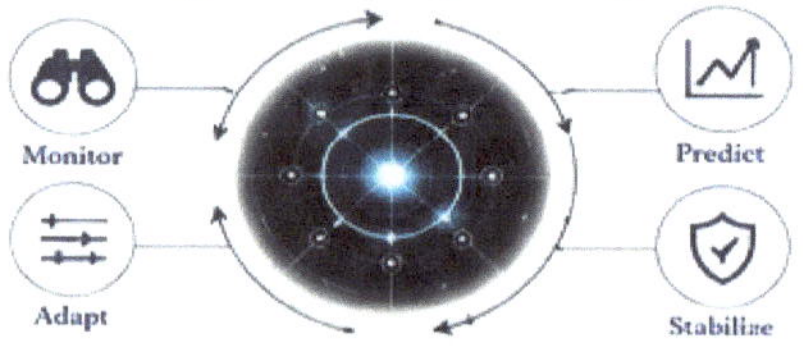

AI-assisted control with physical
limits hard-coded.

10. Civilizational Applications

Energy abundance could enable:

- Interstellar propulsion
- Planetary-scale engineering
- Climate restoration
- Universal knowledge exploration
- Long-term survival of intelligent life

Energy is the foundation of
a flourishing cosmic future.

11. Challenges Ahead

Many hurdles remain:

- Extreme environment materials
- Quantum gravity effects
- Navigation and control precision
- Ethical governance
- Preventing misuse
- Unknown unknowns

Caution, research, and humility
are our best tools.

12. A Vision of Tomorrow

One day, humanity—or our descendants—may
use black hole energy with wisdom and purpose.

Not to dominate the cosmos,
but to participate in it.

The Ultimate Goal

Not unlimited power,
but a deeper partnership
with the Universe.

Remember

We seek not to exploit,
but to understand.
We seek not to conquer,
but to coexist.

"The future is written in possibilities.
Knowledge is the key.
Wisdom is the compass.
The journey has just begun."

The technologies of tomorrow will be built on the physics of today,
guided by ethics, and inspired by wonder.

Future Observatories and Experiments

SEEING THE UNSEEN — TESTING THE RIES, REVEALING TRUTHS

New instruments and experiments will open windows into the extreme universe, allowing us to test the information paradox, measure quantum effects, and explore new physics.

1. Next-Generation Radio Arrays

Sharper imaging of black hole shadows and jet structures.

- Higher resolution (sub-microarcsceond)
- Multi-frequency polarimetry
- Real-time monitoring & variability

Revealing accretion physics, magnetic fields, and horizon-scale structure.

2. Space-Based Gravitational Observatories

Detecting ripples in spacetime from black hole mergers.

- Lower-frequency waves (mHz–Hz)
- Observe supermassive black hole mergers
- Map population across cosmic history

Testing general relativity in the strong-field regime and probing quantum gravity clues.

3. High-Energy Neutrino Telescopes

Neutrinos can escape dense environments and travel across the cosmos.

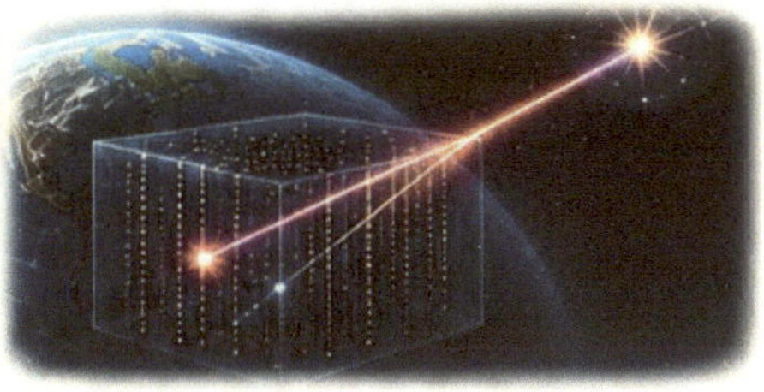

- Detect neutrinos from black hole jets
- Probe hadronic processes
- Test models of energy extraction

A new messenger to unveil hidden engines of extreme energy.

4. X-Ray and Gamma-Ray Observatories

Mapping the hottest and most energetic regions.

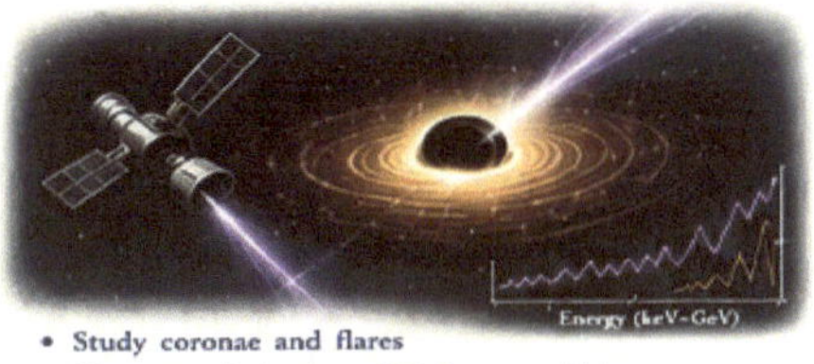

- Study coronae and flares
- Measure polarization of high-energy light
- Search for spectral signatures of new physics

Unlocking the physics of accretion, coronae, and particle acceleration.

5. Quantum Simulation Labs

Recreating black hole analogues in controlled systems.

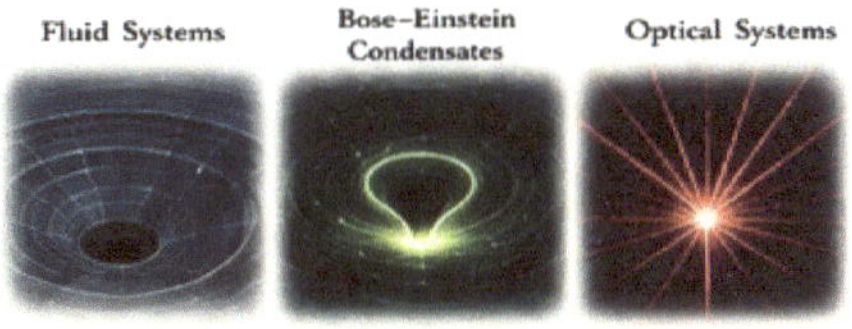

- Event horizon analogues
- Test Hawking-like radiation
- Explore entanglement and information flow

Bringing aspects of black hole physics into the laboratory.

6. Quantum Information Experiments

Probing entanglement, error correction, and information retrieval.

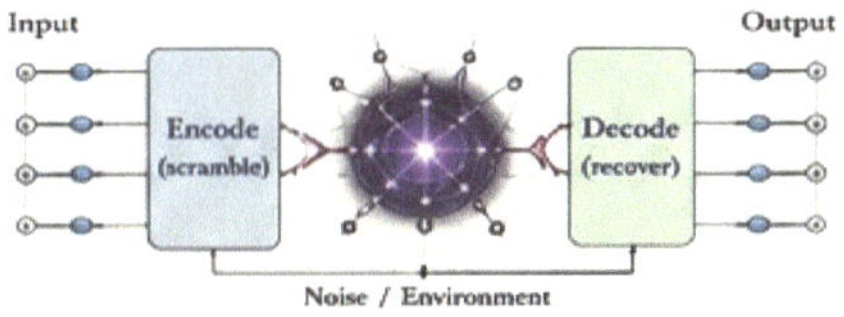

- Test quantum error correction at scale
- Explore black hole codes
- Investigate entanglement structure

Learning how information survives in noisy, curved spacetimes.

7. Pulsar Timing Arrays

Using millisecond pulsars as cosmic clocks to detect gravitational waves.

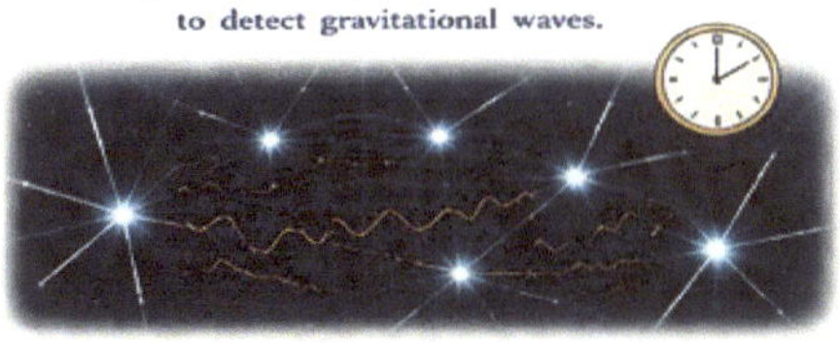

- Monitor many pulsars with extreme precision
- Detect nanohertz gravitational waves
- Probe supermassive black hole binaries & galaxy-scale detector for the lowest-frequency waves.

A galaxy-scale detector for the lowest-frequency waves.

8. Very Large Telescope Networks

High-resolution optics and interferometry for direct imaging.

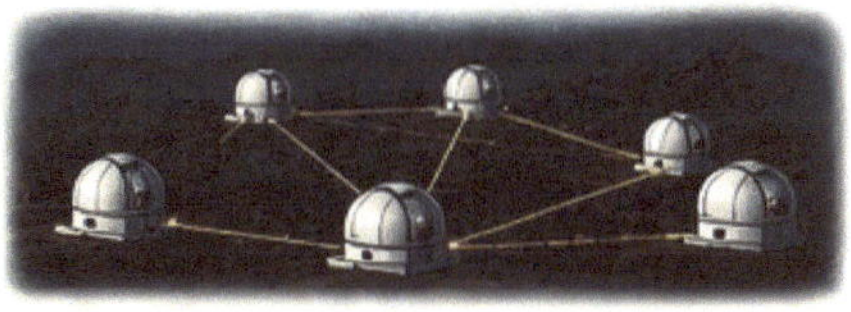

- Extremely large apertures
- Interferometric resolution
- Direct imaging of exoplanets and disks

Seeing the environments where black holes grow and interact.

9. Deep-Space Probes and Starships

Future missions to explore and measure extreme gravitational environments.

- Measure gravitational gradients in situ
- Probe event horizons from close range
- Search for new forms of energy extraction

Humanity's first direct exploration of extreme spacetimes.

10. Cosmic Surveys and 21-cm Maps

Charting matter, dark energy, and the large-scale structure.

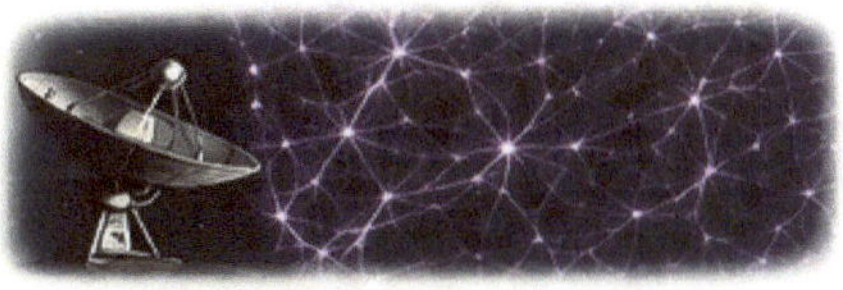

- 21-cm tomography of the early universe
- Map dark matter and baryonic structure
- Constrain cosmological models

Understanding the big picture in which black holes evolve.

11. Theory–Experiment Synergy

Data and theory must evolve together.

- Iterative feedback
- Refine models
- Inspire new experiments

Progress comes from the dance between ideas and observations.

12. The Road Ahead

The next decades will be transformative.

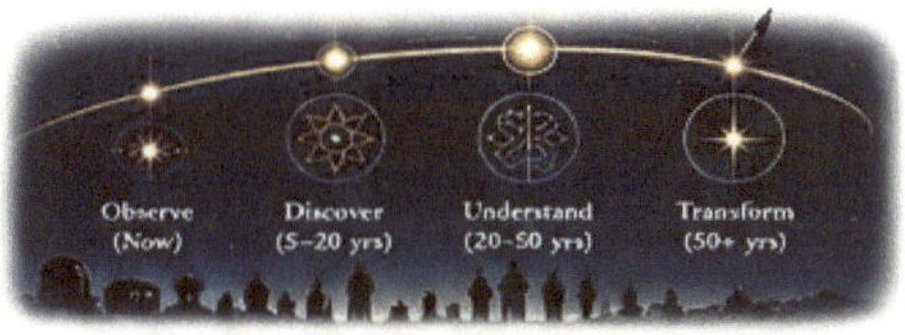

- New tools will reveal new truths
- Surprises are inevitable
- The universe still holds its deepest secrets
- Curiosity, courage, and collaboration will carry us beyond today's horizons.

The technologies of tomorrow will be built on the physics of today, guided by ethics, and inspired by curiosity.

39 THE FUTURE IS OPEN

THE NEXT FRONTIERS IN BLACK HOLE PHYSICS AND ENERGY EXTRACTION

We do not know what lies beyond the horizon—
but we know the questions that will take us there.

1. Key Open Questions

- What replaces the singularity?
- How is information encoded in the final state?
- Can we harness more energy without violating fundamental limits?
- What is the true nature of spacetime at the Planck scale?
- Do all black holes follow the same ultimate fate?
- How do black holes shape the quantum structure of the universe?

2. Emerging Directions

Quantum Gravity Theories
Long quantum gravity, string theory, causal sets, symplectic gravity, and beyond.

Black Hole Information
Islands, wormholes, replica geometries, and quantum error correction.

Energy Extraction Innovations
New mechanisms using plasma physics, quantum fields, and engineered spacetimes.

Multi-Messenger Astronomy
Gravitational waves, neutrinos, and high-energy photons reveal black hole secrets.

3. The Road Ahead: from Theory to Technology

Theoretical Insight

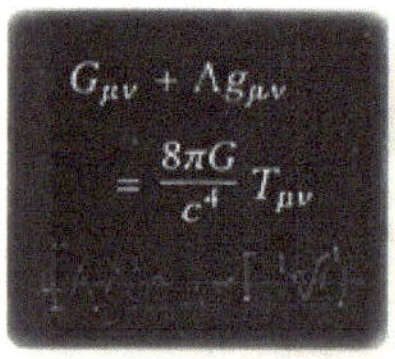

Deeper understanding of spacetime, quantum fields, and information.

→

Mathematical Frameworks

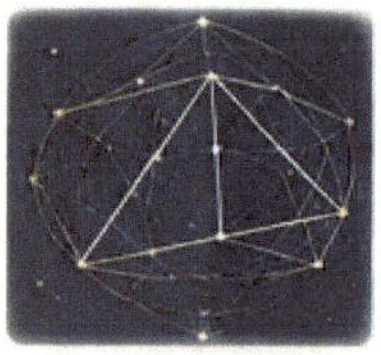

Develop new tools to unify quantum theory and gravity.

→

Observational Tests

Precision data from EHT, LISA, SKA, and future observatories.

→

Engineering Applications

Prototype energy extractors and stabilize extreme environments.

→

Sustainable Utilization

Safe, scalable, and ethical use of energy from the cosmos.

→

Expanding Horizons

New technologies open new frontiers for humanity.

4. Potential Impacts

- Clean, abundant energy for the future
- Interstellar exploration and civilization growth
- Advances in fundamental physics and technology
- A deeper connection between humanity and the universe

5. A Vision of the Future

6. Our Guiding Principles

Curiosity
Ask boldly. Question deeply. Seek truth.

Responsibility
Use knowledge for the benefit of all.

Collaboration
Science knows no borders. Together we go further.

Humility
The universe is vast. We are still learning.

"The horizon is not an end,
but a doorway to discovery.
The future is open—
and we are its explorers."

The Journey Continues

Beyond the Event Horizon, Beyond Our Imagination

The end of this book is not the end of the story — it is an invitation to explore.

What We Have Learned

1. Black Holes Are Real
From theoretical curiosity to observational certainty, black holes are among the most fascinating objects in the universe.

2. They Teach us Extremes
Black holes reveal the limits of gravity, time, space, and the laws of physics themselves.

3. Hawking Radiation Changed Everything
Quantum effects near the horizon showed that black holes are not completely black.

4. Information Matters
The information paradox challenges our deepest understanding of reality and quantum mechanical laws.

5. Energy Can Be Extracted
From rotation, fields, and quantum fluctuations, black holes are not just cosmic sinks — they are cosmic engines.

6. We Are Just Beginning
New theories, new technologies, and new ideas will take us even further.

The Grand Outlook

Unified Physics

Black holes may hold the key to unifying general relativity and quantum mechanics.

Quantum Information

Holography, islands, and error correction point to a deeper structure of reality.

Cosmic Origins

Black holes may illuminate the birth of the universe and the nature of time itself.

Energy Futures

Harnessing black hole energy could one day transform our civilization.

Beyond Our Horizon

Countless mysteries remain, waiting for curious minds to explore.

Our Place

In understanding black holes, we better understand ourselves and our cosmic home.

"Black holes are not eternal prisons as we once thought. They are sources and sinks of energy, and possibly gateways to new understanding."
— *Stephen Hawking*

"The important thing is not just to survive, but to thrive with passion and compassion and humor and style and generosity and kindness."
— *Stephen Hawking*

A Message to the Reader

Science advances through questions that dare to go where no one has gone before. Keep your curiosity alive. Question, imagine, calculate, observe, and dream. The universe is vast, and we are its explorers.

Keep Looking Up.

Key Takeaways

- Black holes connect gravity, quantum mechanics, information, and energy.
- They challenge our theories and expand our imagination.
- Energy extraction is possible — and may shape the future.
- The deepest mysteries inspire the greatest discoveries.
- The universe is still full of wonders.

Beyond the Horizon

Every answer opens the door to new questions.

Every discovery reveals new frontiers.

The journey continues... are you ready?

THANK YOU FOR JOINING THIS JOURNEY.
THE UNIVERSE AWAITS.

"Somewhere, something incredible is waiting to be known."

PREFACE

Penrose Diagrams for the Most Important Spacetimes

— Especially Related to Black Holes —

Roger Penrose introduced conformal diagrams in the 1960s as a revolutionary way to visualize the global causal structure of spacetime. Today, these diagrams—commonly known as Penrose diagrams—are among the most powerful conceptual tools in general relativity. By applying a conformal transformation that brings infinity to a finite boundary while preserving the paths of light rays, Penrose diagrams reveal the complete causal relationships within a spacetime on a single page.

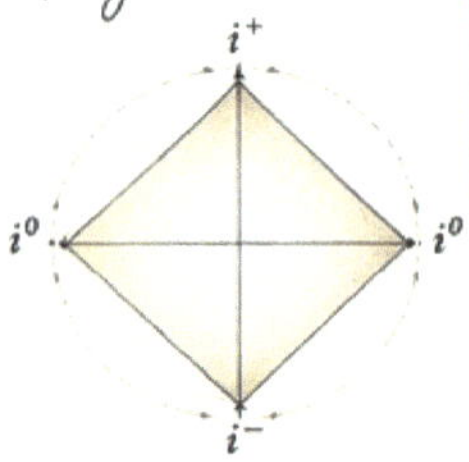

Conformal compactification

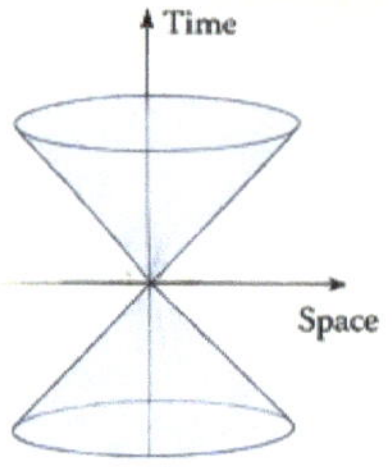

Light cone
what can influence what

Unlike ordinary coordinate plots, Penrose diagrams do not represent physical distances. Instead, they illustrate which events can influence one another, where horizons form, how singularities arise, and what regions of spacetime remain causally disconnected. They allow physicists to study entire universes, black holes, cosmological models, and gravitational collapse within compact geometric figures.

What we see depends on where we are.
What we can see depends on the spacetime we live in.
Penrose diagrams make this visible.

This volume presents a carefully selected collection of Penrose diagrams for many of the most important spacetimes encountered in modern gravitational physics. Beginning with the foundations of conformal compactification and the interpretation of infinities, the reader is guided through progressively richer geometries, including Minkowski spacetime, accelerated observers in Rindler spacetime, Schwarzschild and Kerr black holes, charged Reissner–Nordström solutions, Schwarzschild–de Sitter and Kerr–de Sitter universes, anti–de Sitter spacetime, the BTZ black hole, Vaidya radiating spacetime, eternal black holes, and other geometries of central importance in contemporary research.

Conformal Transformation

$$g_{ab} \rightarrow \tilde{g}_{ab} = \Omega^2 g_{ab}$$

with $\Omega = 0$ at infinity

preserves light cones

$(ds^2 = 0)$

but brings infinity to a finite boundary.

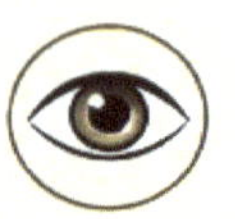

Each page has been designed not merely as an illustration but as a self-contained reference sheet. Every diagram combines the global causal structure with concise explanations of horizons, singularities, conformal boundaries, characteristic regions, important equations, and their physical interpretation. The emphasis throughout is on clarity, consistency, and visual intuition while maintaining scientific accuracy.

Penrose Diagrams play an Indispensable role in many Active Areas of Theoretical Physics

- Classical General Relativity
- Black Hole Thermodynamics
- Hawking Radiation
- Gravitational Collapse
- Cosmic Censorship
- Quantum Field Theory in Curved Spacetime
- AdS/CFT Correspondence
- Quantum Gravity
- Information Paradox Research
- Modern Cosmology

The hope is that this collection will serve not only as an educational atlas but also as a source of inspiration. Behind every Penrose diagram lies a profound story about the geometry of spacetime, the propagation of light, the nature of horizons, and the ultimate limits of observation. These elegant geometric maps remind us that understanding the Universe often begins not with measuring distances, but with understanding causality itself.

May these pages encourage readers to look beyond coordinates and equations, toward the deeper geometric language through which spacetime reveals its structure.

Whether you are a student discovering relativity for the first time, an educator, or a researcher exploring the frontiers of gravity, may these diagrams guide your understanding and ignite your curiosity about the cosmos.

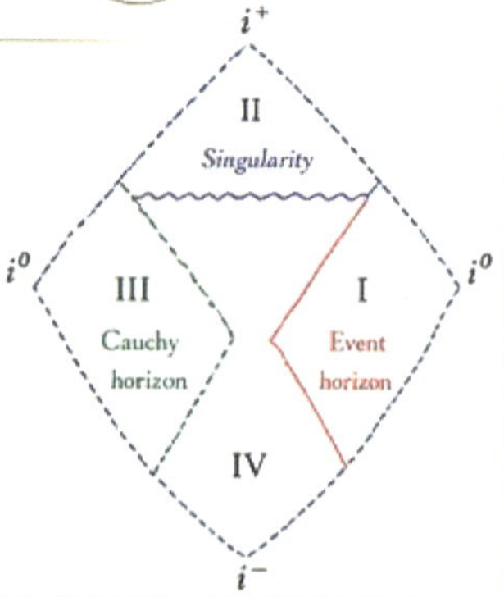

Sandi Setiawan
Life Member, Clare Hall
University of Cambridge

Penrose diagrams are not just pictures. They are maps of possibility and limitation—showing what can influence what, and how the Universe is woven by light, gravity, and fate.

"Black holes are not the end of everything. They are the beginning of new physics and new possibilities."
— Stephen Hawking

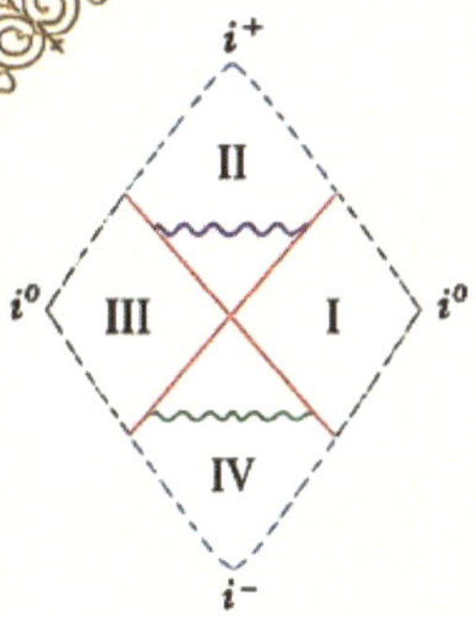

TABLE OF CONTENTS

PENROSE DIAGRAMS FOR IMPORTANT SPACETIMES

This collection presents Penrose diagrams of the most important spacetimes in General Relativity and Modern Physics. Each page is a self-contained reference including the diagram, key features, equations, regions, boundaries, and physical interpretation.

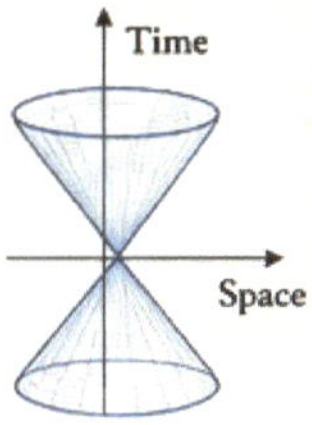

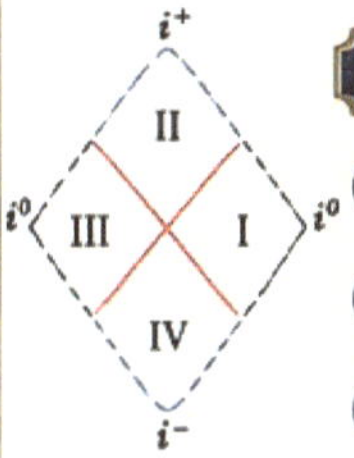

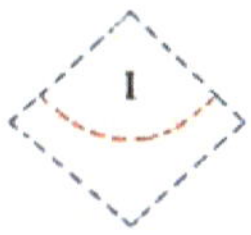

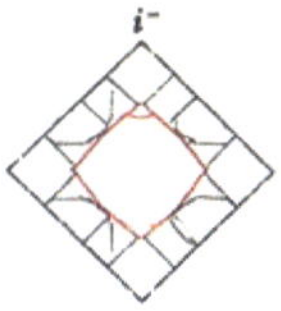

PART I — FOUNDATIONS

PART II — FLAT AND ACCELERATED SPACETIMES

PART III — BLACK HOLE SPACETIMES

PART IV — COSMOLOGICAL SPACETIMES

PART V — BLACK HOLES IN COSMOLOGICAL BACKGROUNDS

PART VI — LOWER-DIMENSIONAL AND DYNAMICAL BLACK HOLES

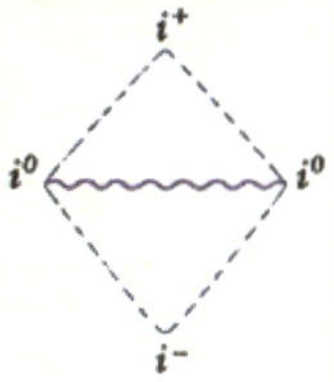

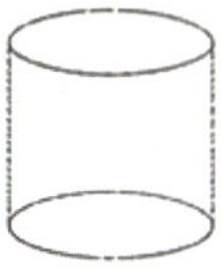

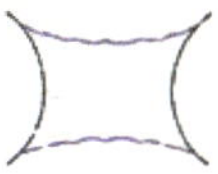

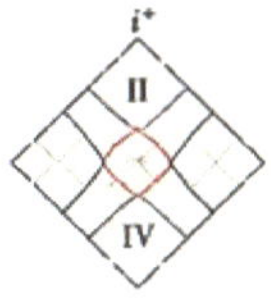

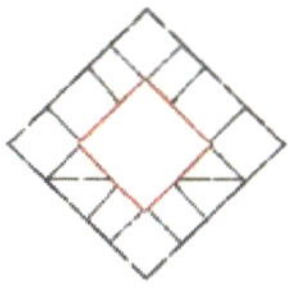

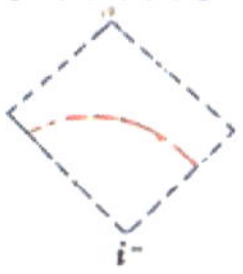

SUMMARY OF ALL PENROSE DIAGRAMS

- ★ A unified comparison of key spacetimes
- ★ Common features, horizons, and boundaries
- ★ Causal structures at a glance
- ★ Similarities and differences
- ★ Quick reference for students and researchers

SYMBOL AND LINE CONVENTIONS

- – – – Null infinity (boundary)
- ∿∿ Curvature singularity ($r = 0$)
- —— Event horizon / Apparent horizon
- ∿∿ Timelike infinity (i^+, i^-)
- —— Null surface (constant r or v)

REGION LABELS (TYPICAL)

- I : Exterior region (observable universe)
- II : Other asymptotic region / exterior
- III : Interior region (inside horizon)
- IV : Black hole interior / trapped region
- V, VI, ... : Additional regions by extension

HOW TO READ A PENROSE DIAGRAM

1. Identify infinities (i^+, i^-, i^0, $\mathscr{I}^-$).
2. Locate horizons (event / apparent).
3. Note singularities (curvature or coordinate).
4. Determine regions (exterior, trapped, etc.).
5. Follow null lines at 45° to understand causal connections.

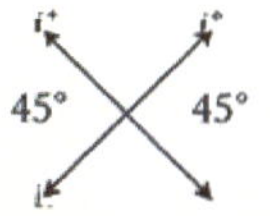

ESSENCE

Penrose diagrams are maps of possibility and limitation—showing what can influence what, and how the universe is woven by light, gravity, and fate.

1 Introduction to Penrose Diagrams

A Map of the Infinite

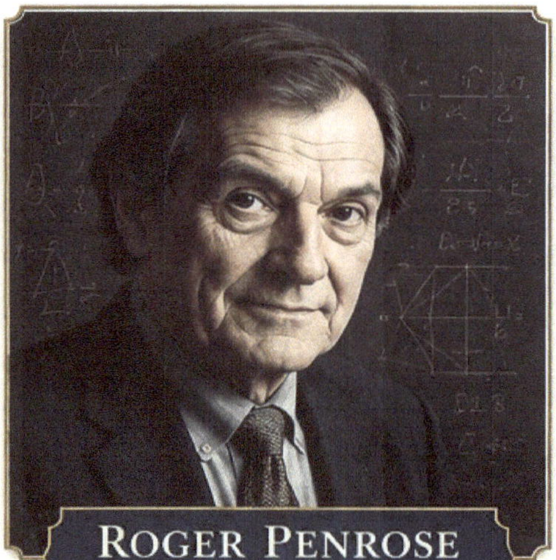

Roger Penrose
1931–

Penrose diagrams are conformal representations of spacetime that capture its causal structure in a finite, elegant way.

In general relativity, spacetime is four–dimensional and often infinite in extent. Yet the essential physics—what can influence what—depends only on causal relationships: the light cones.

Penrose diagrams are conformal (angle-preserving) diagrams that compress infinite regions of spacetime into a finite two–dimensional picture, making the global causal structure clear at a glance.

Why Penrose Diagrams?

- They bring infinity to a finite canvas.
- They preserve causal structure (who can affect whom).
- They reveal horizons, singularities, and trapped regions.
- They unify many spacetimes under one visual language.
- They are powerful tools for intuition and discovery.

The Core Idea

Real Spacetime (Infinite) | **Conformal Compression** | **Penrose Diagram (Finite)**

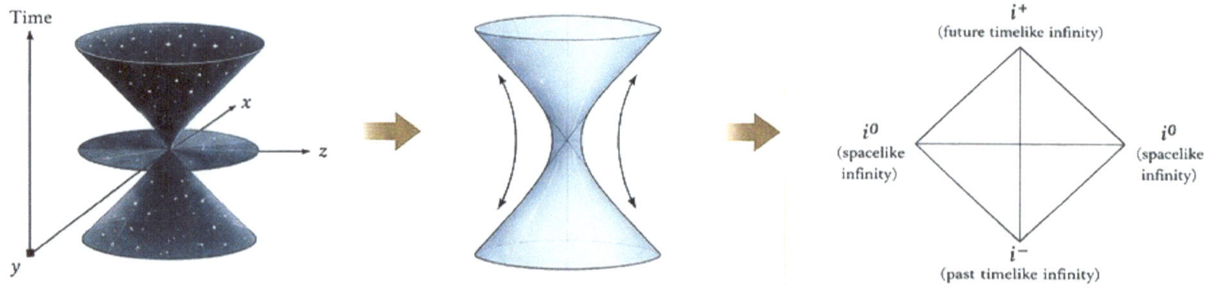

Four-dimensional and typically infinite.

Distances are rescaled, angles (causal structure) preserved.

The entire infinite spacetime in a finite diagram.

What You Can See

- Event horizons (boundaries of no return)
- Singularities (where curvature diverges)
- Regions that can communicate
- The global structure of space and time
- Even the fate of the entire universe

Not Just Pictures—Windows Into Truth

Penrose diagrams do not show distances or durations accurately. They show what matters most: the causal skeleton of reality.

From black holes to expanding universes, Penrose diagrams reveal the deep architecture of spacetime.

> *"The aim is not to describe the universe in detail, but to understand its possibilities."*
> — Roger Penrose

2 How to Read a Penrose Diagram

A Practical Guide

Penrose diagrams look simple, but every line and point has precise meaning. This page explains the symbols, boundaries, and directions you will use throughout this book.

The Building Blocks

Singularity
A place where curvature diverges and known physics breaks down.

Singularity (Spacelike)
A boundary that exists at the same time everywhere.

Event Horizon
A null surface. Nothing inside can send signals to the outside universe.

Other Horizons
Cauchy horizons, cosmological horizons, or inner horizons.

Null Boundary
Lines at 45°. Represents the path of light.

Spacetime Region
The shaded area is a real region of spacetime.

A Typical Diamond-Shaped Diagram

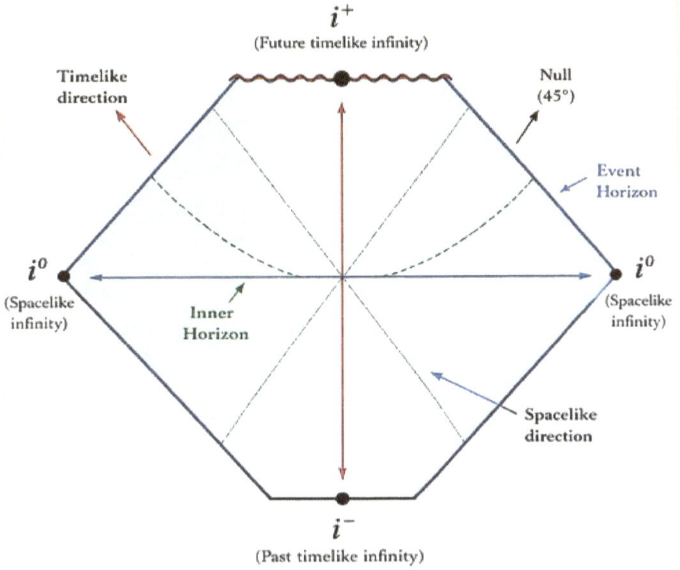

Infinity Points

i^+ Future timelike infinity. The endpoint of all future-directed timelike worldlines.

i^- Past timelike infinity. The starting point of all past-directed timelike worldlines.

i^0 Spacelike infinity. Points approached by traveling infinitely far at one instant of time.

$\mathcal{J}^+$ Future null infinity. Where light rays go that escape to infinity in the future.

$\mathcal{J}^-$ Past null infinity. Where light rays come from that started infinitely far away in the past.

Types of Lines (Causal Character)

Timelike

Inside the light cone.
Slower than light.
Can change your position in time.

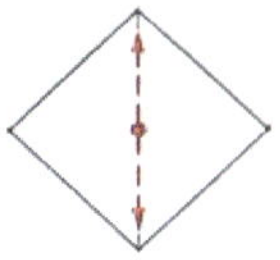

Example:
An observer moving along a worldline.

Null (Lightlike)

On the light cone (45° lines).
Travel at the speed of light.
Separates inside from outside.

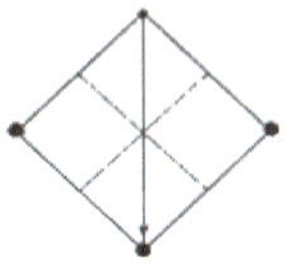

Example:
A photon worldline.

Spacelike

Outside the light cone.
Faster than light (not possible for matter).
Represents "now" at one instant.

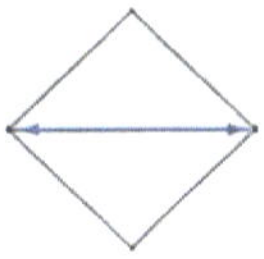

Example:
A snapshot of the universe at one time.

Tips for Reading

- Follow the arrows. Up is future, down is past.
- 45° lines are light rays. Nothing can go outside a horizon.
- Boundaries matter. Horizons tell you what you can see or influence.
- Regions are real. Each shaded area is an actual part of the spacetime.
- The whole picture. Together, the diagram shows the complete causal structure.

Conformal Magic

Penrose diagrams are made using a conformal transformation, which keeps angles (causal structure) the same but rescales distances so that infinity is brought to a finite boundary.

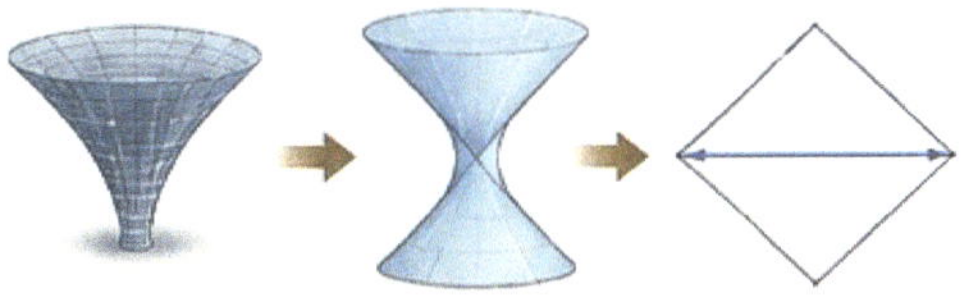

Real spacetime — Conformal map (rescale + compactify) — Penrose diagram (finite)

> *"The power of Penrose diagrams lies in their ability to show the infinite with the simplicity of a picture."*
> — Roger Penrose

What They Tell Us

- Causal structure of the universe
- Visibility and horizons
- Black holes and singularities
- Gravitational collapse
- Cosmological evolution
- And much more...

Master the Diagrams,
Master the Universe.

3 MINKOWSKI SPACETIME

(FLAT SPACETIME)

Minkowski spacetime is the simplest solution of general relativity: flat, infinite, and without gravity.
It is the background for special relativity and the starting point for understanding all other spacetimes.

KEY FEATURES

- No gravity: spacetime is flat.
- Infinite in all directions.
- No curvature.
- Light travels on 45° lines.
- Causal structure is simple and global.
- Serves as the reference for all other spacetimes.

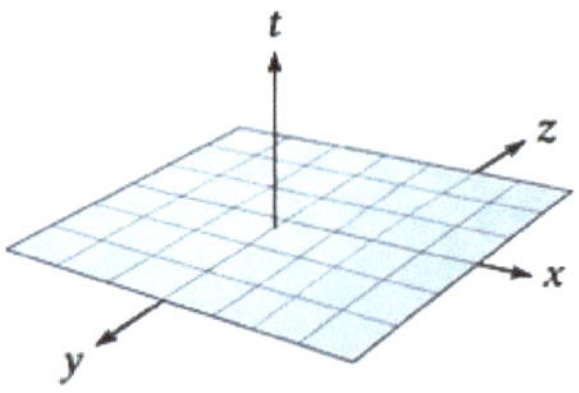

PENROSE DIAGRAM

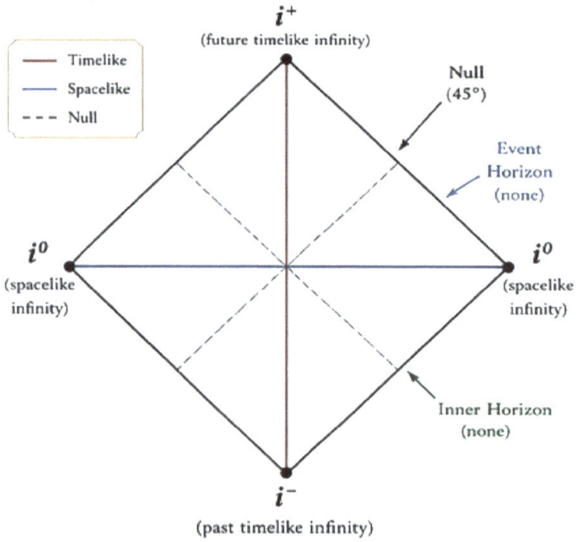

WHAT IT REPRESENTS

- An infinite, empty universe.
- No black holes, no horizons.
- Light cones everywhere look the same.
- Basis for defining causality.
- The simplest stage on which physics unfolds.

INTERPRETING THE DIAGRAM

i^+ Future timelike infinity
Where all future-directed timelike observers end up (arbitrarily far in the future).

i^- Past timelike infinity
Where all past-directed timelike observers come from.

i^0 Spacelike infinity
Points that are reached at a single instant of time (arbitrarily far away).

Null (45°) lines
Paths of light rays.
They separate the diagram into causal regions.

Timelike lines
Paths of objects moving slower than light. Cannot cross the 45° boundaries to reach outside.

Spacelike lines
Paths that would require moving faster than light. Connect points that cannot influence each other.

Shaded region
Represents all of Minkowski spacetime (infinite), compressed into a finite diagram.

PHYSICAL MEANING

If an event lies outside your light cone, no signal can reach you from it (or to it).
Causality is absolute.

FROM REAL SPACETIME TO PENROSE DIAGRAM

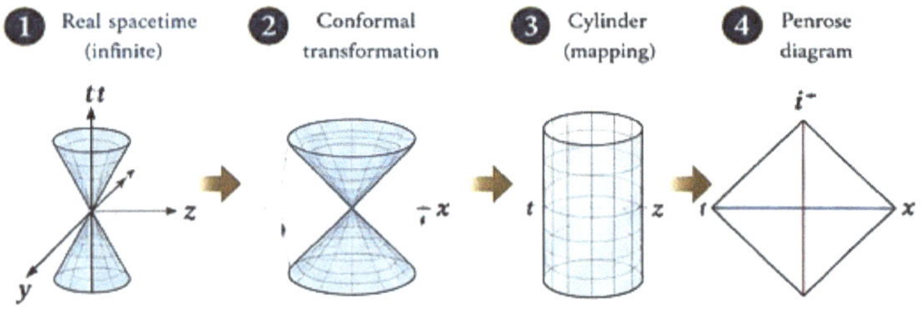

Infinite in extent.

Rescale distances so that infinity is brought to a finite boundary.

Map the infinite space onto a finite cylinder.

Flatten the cylinder into a finite diamond. Causal structure is preserved.

CAUSAL REGIONS

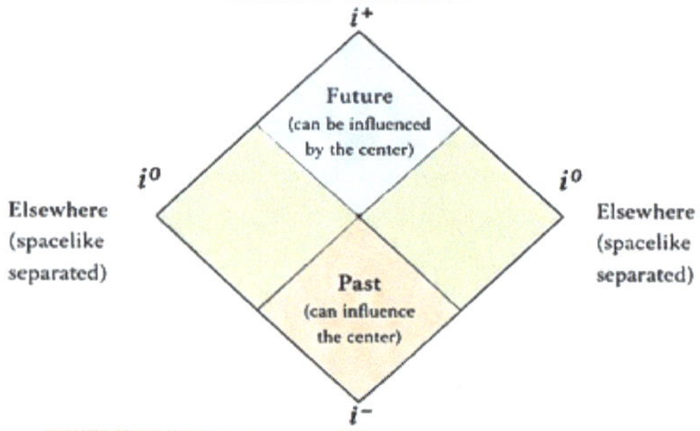

Events in the green regions cannot affect or be affected by the central event — they are spacelike separated.

WHY IT MATTERS

- Defines light cones and causality.
- Foundation for all relativistic physics.
- Helps compare curved spacetime to flat.
- Essential for understanding horizons and singularities.
- A map that turns infinity into insight.

4 2D Minkowski Spacetime

The Causal Diamond

The 2D Minkowski diagram is the simplest possible Penrose diagram. It captures all the essential causal structure of spacetime in two dimensions.

Key Features

- Two dimensions: time (t) and one space (x).
- Flat spacetime (no gravity).
- Light travels at 45°.
- Perfect symmetry.
- The building block for understanding all other spacetimes.

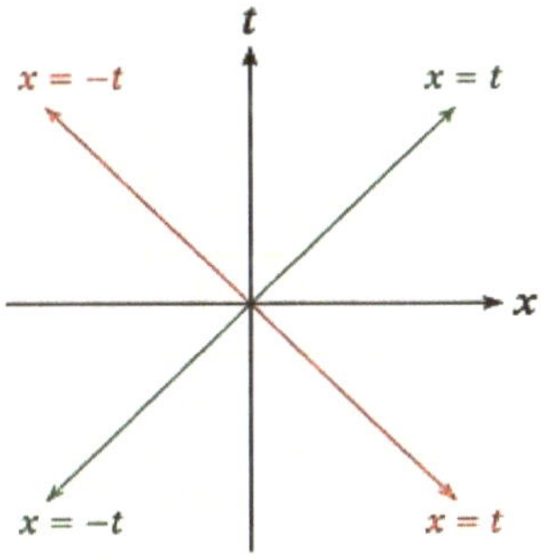

Penrose Diagram

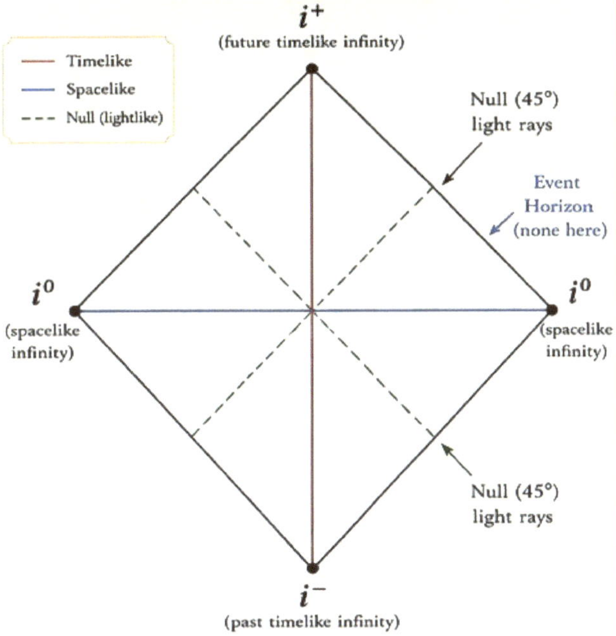

Interpretation

- i^+ Future timelike infinity: events arbitrarily far in the future.
- i^- Past timelike infinity: events arbitrarily far in the past.
- i^0 Spacelike infinity: events arbitrarily far away in space.
- No horizons or singularities. The entire spacetime is visible and accessible.
- Every event can influence and be influenced by others inside its light cone.

Physical Picture

Imagine you are at the center.

- The 45° lines are the paths of light rays.
- Nothing can travel outside the light cone.
- The diagram shows the entire universe—past, present, and future—in a finite picture.

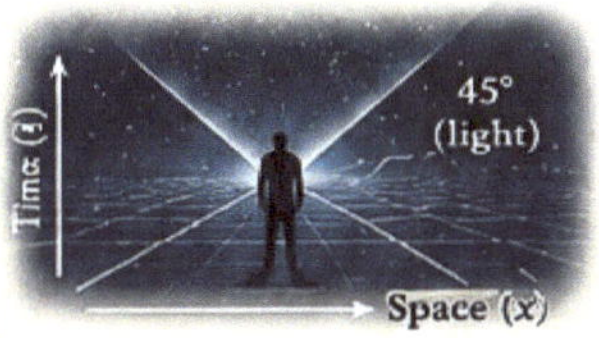

Inside the Diamond

Timelike paths:
Stay within the light cone. You can go to the future or the past, but never outside the cone.

Spacelike paths:
Cross the diagram horizontally. These would require faster-than-light travel.

Null paths (light):
Run along the edges of the diamond at 45°.

All possible events of this 2D universe are contained within this finite diagram.

Coordinate Chart

Real coordinates (t, x):

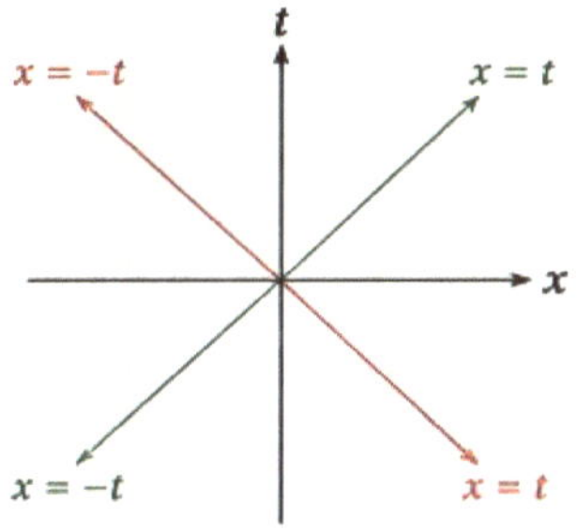

Penrose diagram (conformal map):

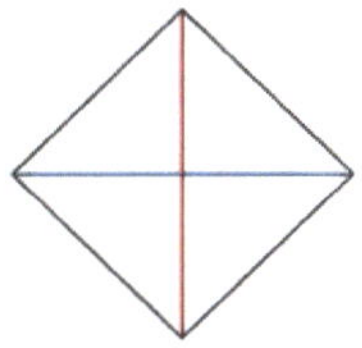

Radial Light Paths

Outgoing Light Rays ($x = \pm t$)

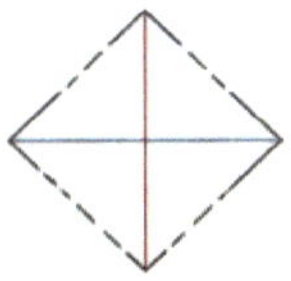

Leave the origin and reach i^0 (spacelike infinity) at time $t = |x|$.

Ingoing Light Rays ($x = \mp t$)

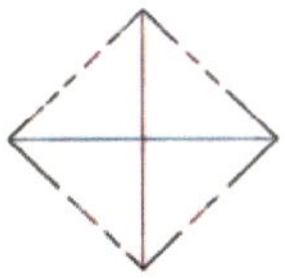

Come from i^0 (spacelike infinity) and reach the origin at time $t = |x|$.

Light rays form the boundaries of the causal diamond.

Causal Structure

The causal diamond shows:

- What can happen.
- What can be seen.
- What can influence you.
- What you can influence.

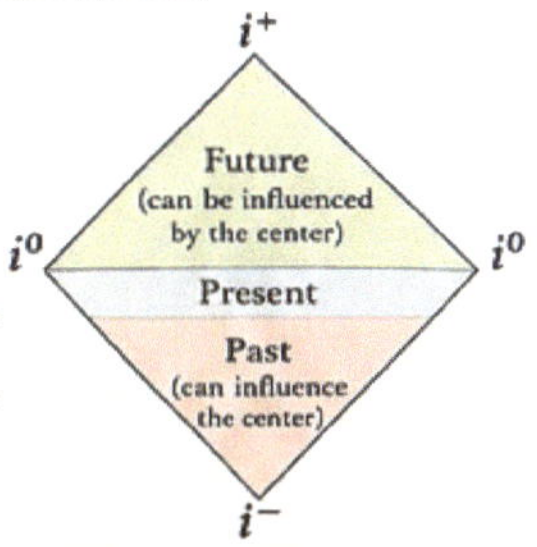

Events outside the light cone are causally disconnected.

Why It Matters

- Simplest example of a Penrose diagram.
- Shows how infinity can be brought to a finite boundary.
- Teaches the meaning of causality.
- Foundation for curved spacetimes and black hole diagrams.
- Essential for understanding cosmology and horizons.

Key Takeaway

The 2D Minkowski diagram is the purest expression of causality in spacetime—a perfect diamond where light defines the boundary between what can, and cannot, affect you.

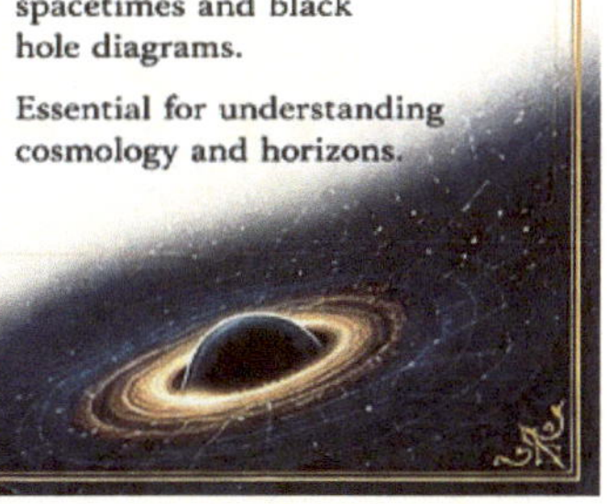

5 Rindler Spacetime

Uniformly Accelerated Observer

Rindler spacetime describes the view of an observer undergoing constant proper acceleration in flat Minkowski spacetime. It reveals horizons even in the absence of gravity.

Key Features

- Flat spacetime, but viewed by an accelerated observer.
- Spacetime is divided into four regions.
- Rindler horizons limit what the observer can see.
- Demonstrates how horizons arise from acceleration alone.
- Prelude to black hole horizons.

Penrose Diagram

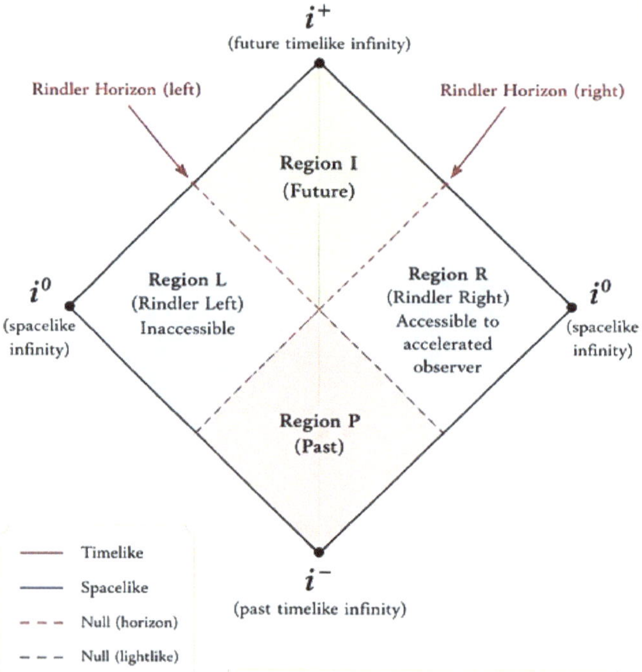

The right wedge (Region R) is the region accessible to a uniformly accelerated observer in the right direction.

Interpretation

- i^+ — Events in Region I are in the observer's future.
- i^- — Events in Region P are in the observer's past.
- i^0 — Spacelike infinity in left and right directions.
- The red dashed lines are Rindler horizons.
- The observer in Region R cannot see Region L.
- Acceleration creates horizons.

Physical Picture

An observer with constant acceleration moves along a hyperbola in Minkowski space.

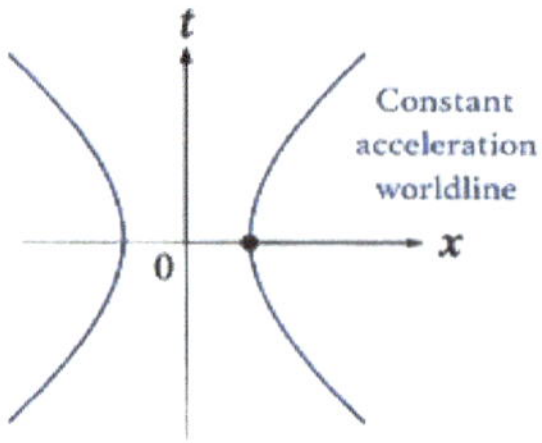

This observer can never receive signals from behind the horizon.

Hyperbolic Worldlines

Worldlines of observers with different constant accelerations.

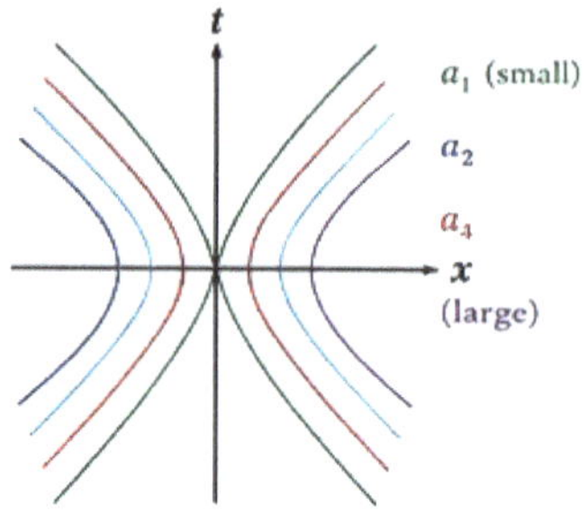

Larger acceleration → closer to the horizon.

Rindler Coordinates

The transformation to Rindler coordinates (t, ξ):

$$t = \xi \sinh(a\eta)$$
$$x = \xi \cosh(a\eta)$$

where a is the proper acceleration.

The observer is at constant $\xi = \frac{1}{a}$.

The Rindler Horizon

The horizons are the null lines:

$$t = \pm x$$

- No signal from beyond the horizon can reach the accelerated observer.
- The horizon is not a physical barrier, but a limitation of the observer's motion.

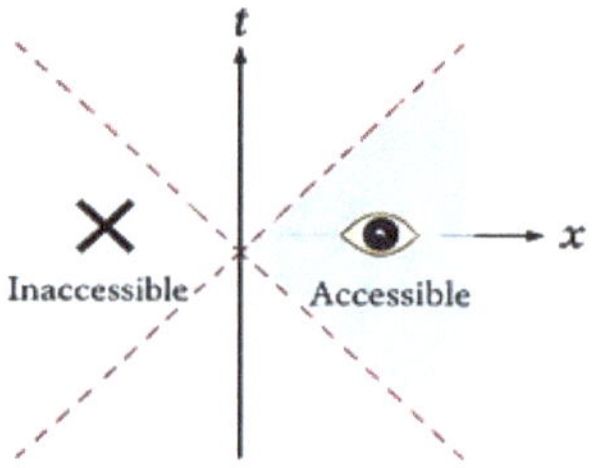

What It Means

- Horizons can arise without gravity, from acceleration.
- This helps us understand black hole horizons as a limiting surface.
- Leads to the concept of Unruh radiation: accelerated observers detect a thermal bath in vacuum.
- Deep connection between acceleration, horizons, and thermodynamics.

Relation to Other Spacetimes

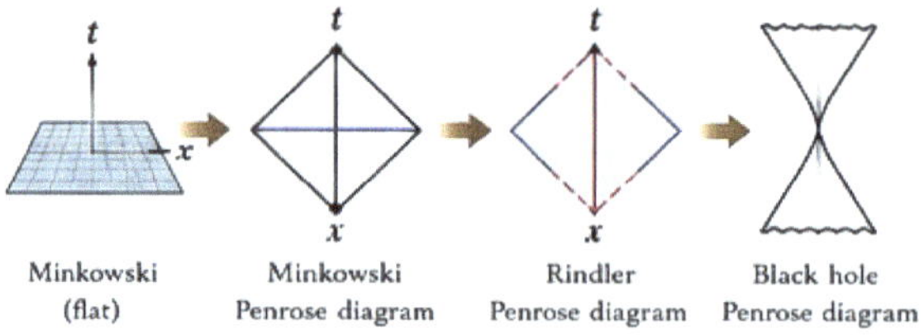

Rindler spacetime is the simplest example showing horizons and causal disconnection in flat spacetime.

Summary

- Flat spacetime.
- Uniform acceleration.
- Horizon appears.
- Limited visibility.
- Foundation for deeper understanding of black holes.

Key Takeaway

Acceleration alone can create horizons. Rindler spacetime shows that what we can observe depends not only on the universe, but also on our state of motion.

6 SCHWARZSCHILD BLACK HOLE

(EXTERIOR REGION)

The Schwarzschild solution ($a = 0$, $Q = 0$) describes a non-rotating, uncharged black hole.
This diagram shows only the exterior region ($r > 2M$) of the full spacetime.

KEY FEATURES

- Spherically symmetric and static.
- Mass M is the only parameter.
- Event horizon at $r = 2M$.
- No access to $r < 2M$ (singularity) from the exterior.
- There is only one asymptotically flat universe (our universe).
- Time runs upward toward the future; space runs rightward.

HORIZON RADIUS

For Schwarzschild black hole,

$$r_s = 2M$$

where $G = c = 1$.

Event horizon: $r = r_s = 2M$

PENROSE DIAGRAM (EXTERIOR REGION)

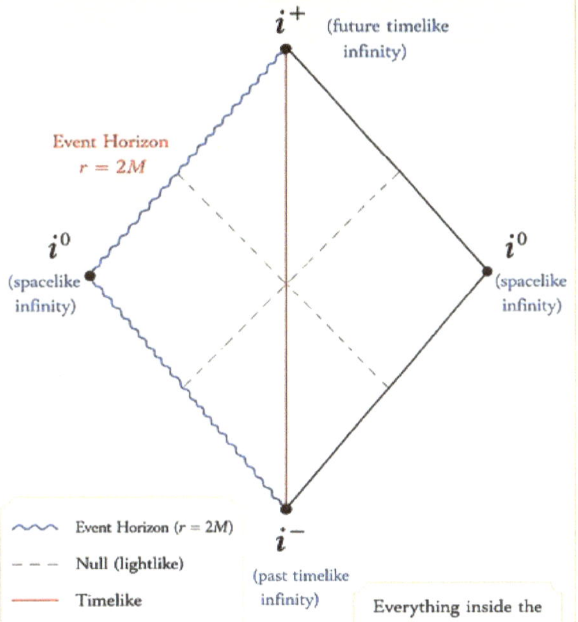

- Event Horizon ($r = 2M$)
- Null (lightlike)
- Timelike
- Timelike (vertical)
- Spacelike

Everything inside the event horizon lies in another region that is not shown here.

INTERPRETATION

- i^+ — Future timelike infinity: where observers go arbitrarily far in time.
- i^- — Past timelike infinity: where events come from arbitrarily far in the past.
- i^0 — Spacelike infinity: points at spatial infinity at one time.
- Event horizon ($r = 2M$): a one-way boundary; cannot be crossed outward.
- Interior ($r < 2M$): not reachable from the exterior region.

KEY FACTS

- No singularity is visible in this diagram.
- Light rays emitted outside can reach i^+.
- Light rays trying to escape near the horizon asymptotically approach it but never cross.
- This region is asymptotically flat.

GEODESIC BEHAVIOR (EXTERIOR REGION)

RADIAL NULL GEODESICS

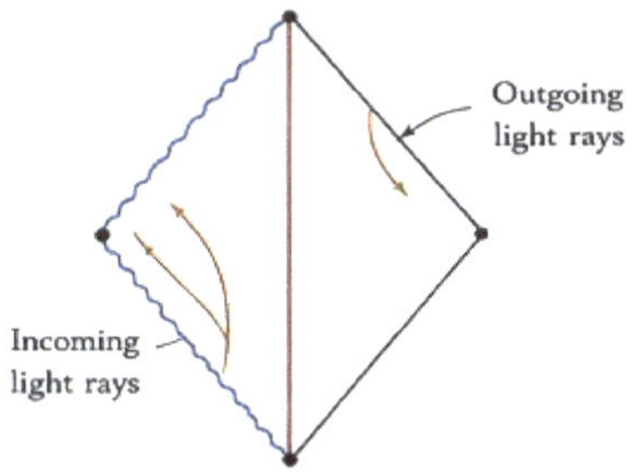

TIMELIKE GEODESICS

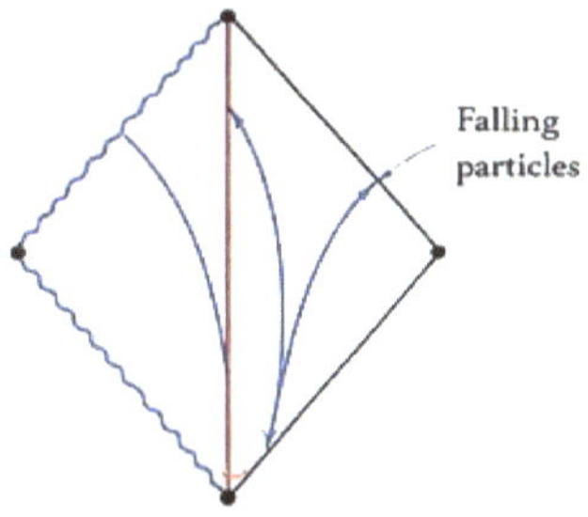

CIRCULAR ORBITS

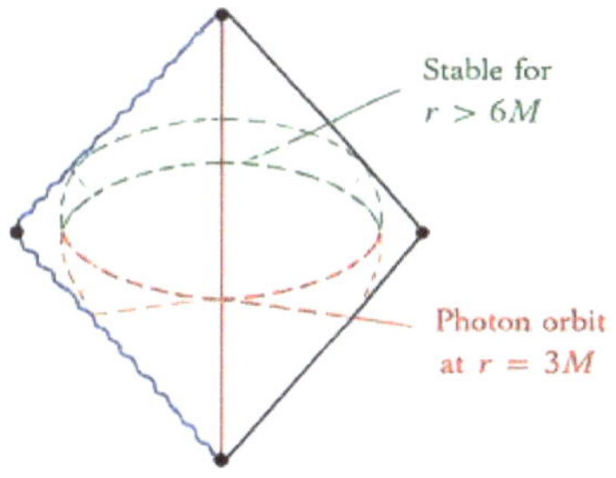

PHYSICAL PICTURE

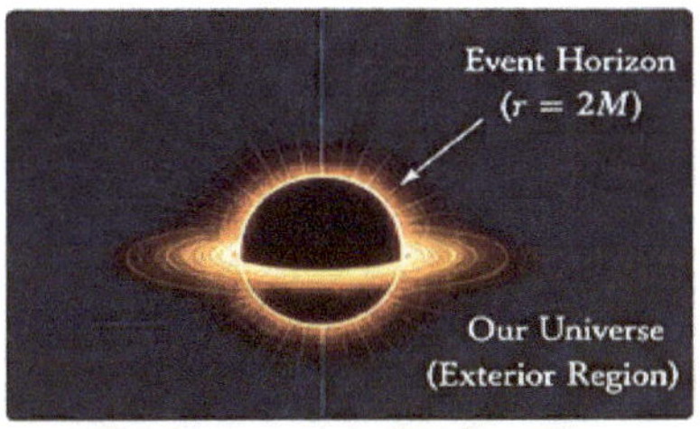

Light can orbit close to the horizon, but nothing can escape once it crosses $r = 2M$.

WHY IT MATTERS

- Foundation of black hole physics.
- Remains relevant in many exact solutions and perturbations.
- Basis for understanding horizons, causality, and information flow.
- Simplest black hole: essential stepping stone to rotating and charged cases.

AT A GLANCE

Parameters	$a = 0$, $Q = 0$, M
Event Horizon	$r = 2M$
Singularity	$r = 0$ (hidden inside)
Spacetime	Static, spherically symmetric
Asymptotic Behavior	Asymptotically flat
Accessible Region (Diagram)	$r > 2M$ only

KEY TAKEAWAY

The Schwarzschild exterior region shows how spacetime is curved around a non-rotating, uncharged black hole. The horizon at $r = 2M$ acts as a point of no return, separating what can be seen from what cannot be reached.

7 SCHWARZSCHILD BLACK HOLE

(MAXIMAL EXTENSION)

Extending the Schwarzschild solution across the horizons reveals an infinite sequence of regions and universes. The full spacetime is symmetric and contains no singularities in the diagram itself.

KEY FEATURES

- The maximal extension contains infinitely many regions (*I*, *II*, *III*, *IV*, ...).
- Universes alternate: some are our own type (black hole exterior), others are "white hole" or other universes.
- The singularity at $r = 0$ is a horizontal line (in the limit of infinite proper time).
- Event horizons ($r = 2M$) are 45° lines in the diagram.
- Crossing a horizon leads to a new region with different physical meaning.
- The diagram is conformal: it preserves causal structure but not distances.

HORIZONS

For Schwarzschild black hole,

$$r_s = 2M$$

Event horizons: $r = r_s = 2M$

Cauchy horizons do not exist in Schwarzschild spacetime.

LEGEND

- $r = 2M$ (Event Horizon)
- $r = 2M$ (Cauchy Horizon — does not exist here)
- $r = 0$ (Singularity)
- Timelike
- Spacelike
- Null (Lightlike)

PENROSE DIAGRAM (MAXIMAL EXTENSION)

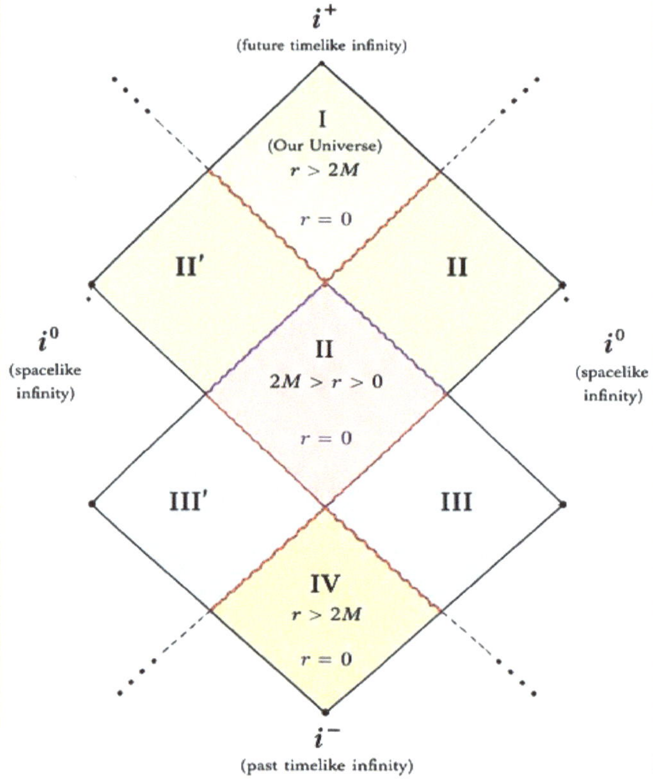

Crossing the event horizon ($r = 2M$) inevitably leads to regions from which one cannot return to our universe (Region I).

INTERPRETATION

- i^+ — Future timelike infinity: where all future-directed timelike paths end.
- i^- — Past timelike infinity: where all past-directed timelike paths begin.
- i^0 — Spacelike infinity: points at spatial infinity at any instant.
- Event horizon ($r = 2M$): one-way boundary; no return once crossed inward.
- Singularity ($r = 0$): a line where curvature diverges.
- The pattern continues infinitely in all directions.

WHAT WE LEARN

- Our universe (Region I) is just one of infinitely many.
- Black holes connect to other regions and other universes.
- The symmetry reflects the time-reversibility of the equations.
- Although singularities exist at $r = 0$, they are hidden behind event horizons from us.

REGION SUMMARY

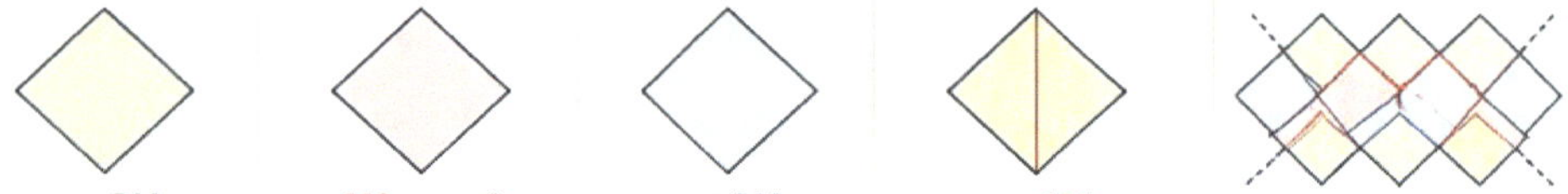

I (Our Universe)	II (Black Hole Interior)	III (Other Universe)	IV (Other Universe Exterior)	... and repeating
$r > 2M$	$2M > r > 0$	$r < 2M$	$r > 2M$	
Exterior region of our universe. Static observers can exist here.	Inside the black hole. All future paths lead to the singularity.	Expanding universe on the other side of the black hole.	Another exterior universe, reachable from Region III but not from I.	The pattern of regions continues infinitely.

CAUSAL STRUCTURE

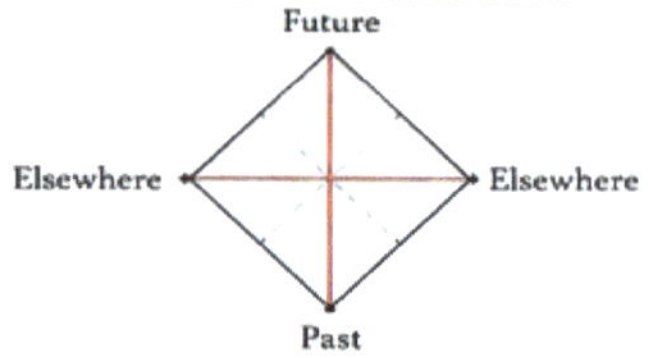

- Timelike: within the light cone
- Null: on the light cone
- Spacelike: outside the light cone

PHYSICAL PICTURE

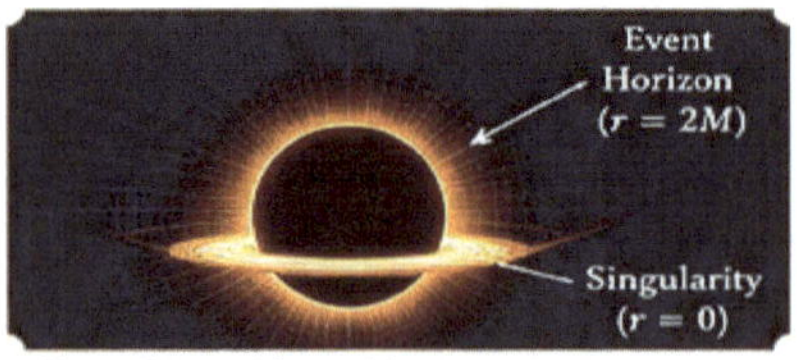

From the outside, nothing can escape once it crosses the event horizon. The black hole connects to other regions and universes.

LIGHT CONES IN DIFFERENT REGIONS

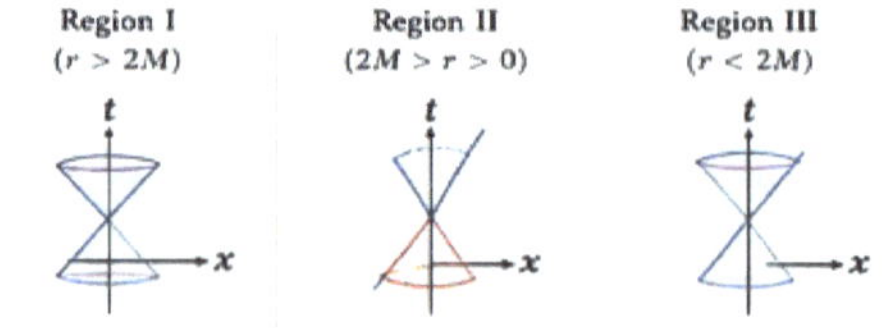

Region I ($r > 2M$)	Region II ($2M > r > 0$)	Region III ($r < 2M$)
Time direction upward.	Time direction becomes inward.	Time direction points outward again.

IMPORTANT POINTS

- Infinite extension: endlessly repeating structure.
- No preferred center: all regions are equivalent.
- Causality is preserved: no paradoxes within classical GR.
- Essential tool for understanding black holes, wormholes, and the nature of spacetime.

KEY TAKEAWAY

The maximal extension reveals that a black hole is not an endpoint of spacetime, but a gateway to other regions and universes, connected through horizons.

8 REISSNER–NORDSTRÖM

(NON-ROTATING CHARGED BLACK HOLE)

The Reissner–Nordström solution describes a non-rotating black hole with mass M and electric charge Q. Depending on Q, the spacetime contains two, one, or no horizons.

KEY FEATURES

- Spherically symmetric, electrostatically charged, non-rotating solution.
- Two parameters: mass M and charge Q.
- Horizons depend on the ratio Q/M.
- When $Q = 0$, reduces to Schwarzschild black hole.
- Maximally charged case has a degenerate horizon.
- Overcharged case has no horizon (naked singularity).

PENROSE DIAGRAM (GENERAL CASE $|Q| < M$)

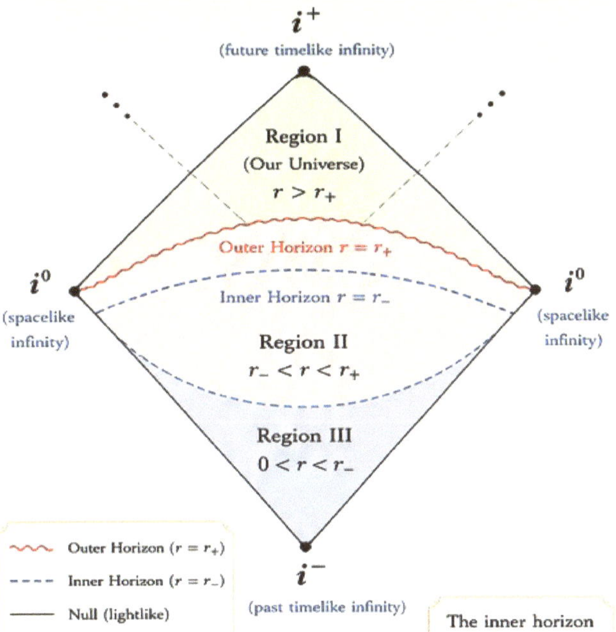

- Outer Horizon ($r = r_+$)
- Inner Horizon ($r = r_-$)
- Null (lightlike)
- Timelike
- Timelike (vertical)
- Spacelike

The inner horizon ($r = r_-$) is unstable to small perturbations (mass inflation).

INTERPRETATION

- i^+ Future timelike infinity: where all future-directed timelike paths end.
- i^- Past timelike infinity: where all past-directed timelike paths begin.
- i^0 Spacelike infinity: points at spatial infinity at one instant.
- Event horizon ($r = r_+$): one-way boundary; nothing can escape outward once crossed inward.
- Inner horizon ($r = r_-$): boundary of Cauchy horizon; instabilities expected.
- Timelike boundaries: separate causally disconnected regions.

HORIZON RADII

The horizon radii are

$$r_\pm = M \pm \sqrt{M^2 - Q^2}$$

Existence requires $|Q| \le M$.

- $|Q| < M$: two horizons ($r_+ > r_- > 0$)
- $|Q| = M$: one degenerate horizon ($r_+ = r_- = M$)
- $|Q| > M$: no horizons

HORIZON SUMMARY

Case	Condition	Horizons
Subextremal	$\|Q\| < M$	Two: r_+, r_-
Extremal	$\|Q\| = M$	One: $r_+ = r_- = M$
Overcharged	$\|Q\| > M$	None

THREE POSSIBILITIES

SUBEXTREMAL ($|Q| < M$)

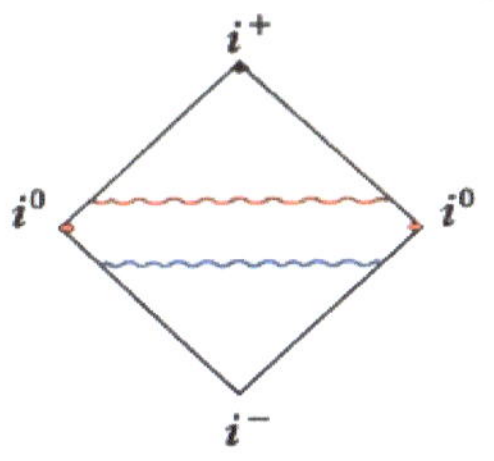

- Two horizons: $r_+ > r_- > 0$
- Cauchy horizon ($r = r_-$) inside
- Four regions (I–IV) connected through the horizons

EXTREMAL ($|Q| = M$)

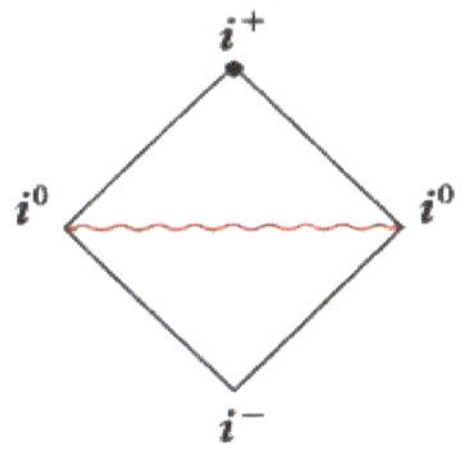

- Degenerate horizon: $r_+ = r_- = M$
- Only two regions (I and II)
- The horizon is both outer and inner

OVERCHARGED ($|Q| > M$)

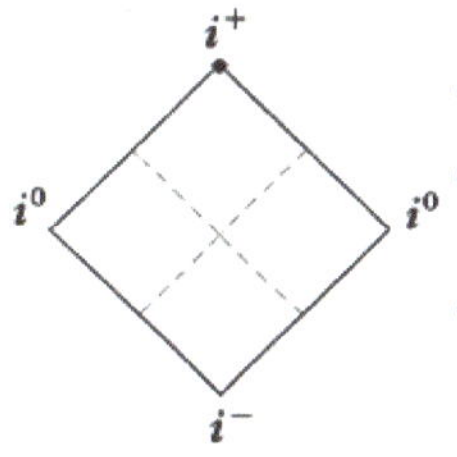

- No real horizons
- Singularity at $r = 0$ is naked
- Not expected in astrophysical black holes (cosmic censorship)

PHYSICAL PICTURE ($|Q| < M$)

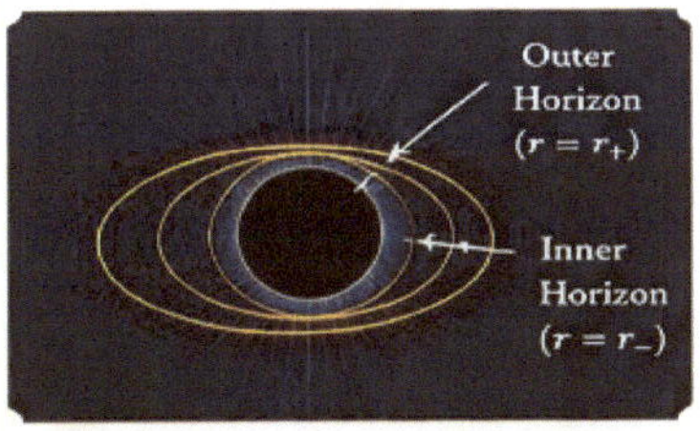

Charge creates an inner horizon and electromagnetic field lines outside. Both horizons trap light and matter.

ELECTRIC FIELD

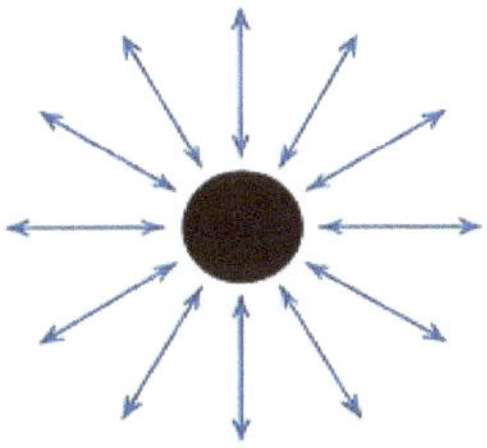

Electric field of a charged black hole points radially outward (or inward for negative charge).

MASS INFLATION

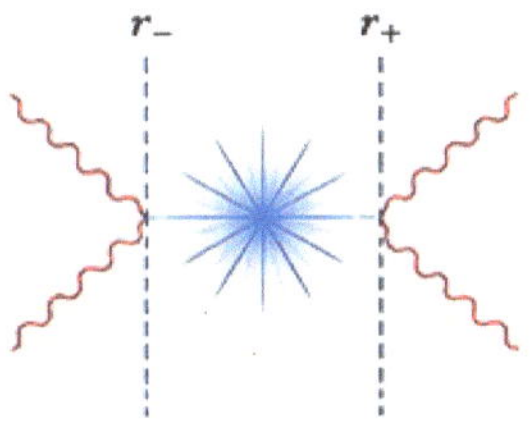

Perturbations crossing the inner horizon (r_-) can experience unbounded growth of mass-energy.

IMPORTANT POINTS

- Infinite extension: endlessly repeating structure.
- Inner horizon is expected to be unstable (mass inflation).
- Reissner–Nordström is the simplest charged black hole.
- Reduces to Schwarzschild when $Q = 0$.
- Cosmic censorship conjecture avoids $|Q| > M$ cases.

KEY TAKEAWAY

The Reissner–Nordström solution shows how electric charge creates inner structure in black holes: two horizons, a Cauchy horizon, and new physical phenomena.

9 EXTREMAL REISSNER–NORDSTRÖM
($Q = M$)

The extremal Reissner–Nordström black hole has charge equal to mass ($|Q| = M$).
The two horizons coincide, creating a degenerate horizon.

KEY FEATURES

- Mass and charge are equal: $|Q| = M$.
- The inner and outer horizons coincide: $r_+ = r_- = M$.
- The horizon is degenerate (double root).
- Surface gravity is zero: $\kappa = 0$.
- The black hole is in a marginally stable state.
- Extremal limit separates subextremal and overcharged cases.

HORIZON RADIUS

For Reissner–Nordström,

$$r_\pm = M \pm \sqrt{M^2 - Q^2}$$

Extremal case: $|Q| = M$
$\Rightarrow r_+ = r_- = M$

LEGEND

- Degenerate Horizon ($r_+ = r_- = M$)
- Would-be Inner Horizon (coincides)
- Null (lightlike)
- Timelike
- Timelike (vertical)
- Spacelike

PENROSE DIAGRAM (EXTREMAL CASE $|Q| = M$)

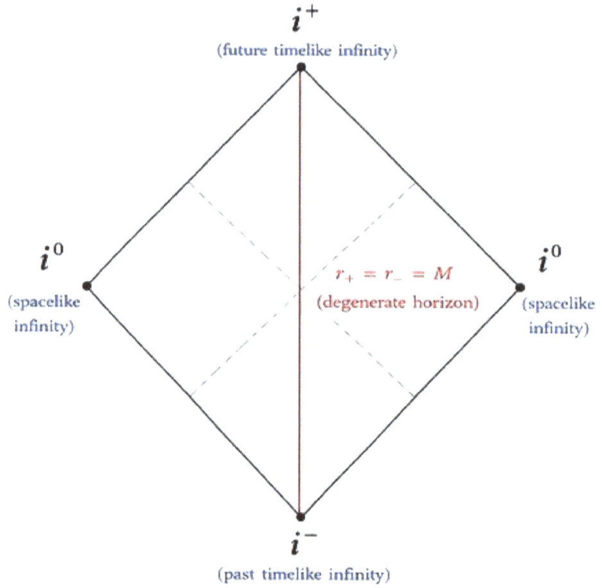

The inner and outer horizons coincide at $r = M$.
No region lies between horizons; the causal structure collapses to a single null surface.

INTERPRETATION

- i^+ Future timelike infinity: where future-directed timelike paths end.
- i^- Past timelike infinity: where past-directed timelike paths begin.
- i^0 Spacelike infinity: points at spatial infinity at one instant.
- Degenerate horizon ($r = M$): one null surface with zero surface gravity.
- No inner region: $r < M$ is not accessible (collapsed).
- Extremal black hole is marginally stable: infinitely cold ($T = 0$).

HORIZON PROPERTIES

Outer Horizon	$r_+ = M$
Inner Horizon	$r_- = M$
Surface Gravity (κ)	0
Hawking Temperature (T_H)	0
Area	$A = 4\pi M^2$
Extremal Limit	$\|Q\| = M$

COMPARISON WITH OTHER CASES

SUBEXTREMAL ($|Q| < M$)

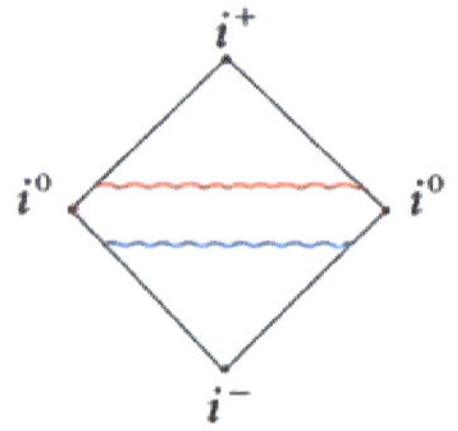

- Two distinct horizons: $r_+ > r_- > 0$.
- Region exists between horizons.
- Non-zero surface gravity ($\kappa > 0$).

EXTREMAL ($|Q| = M$)

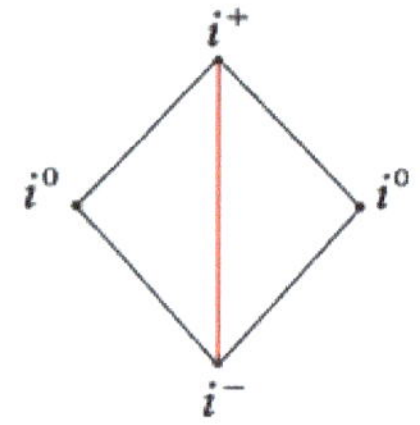

- Horizons coincide: $r_+ = r_- = M$
- No region between horizons.
- Zero surface gravity ($\kappa = 0$).
- Infinitely cold ($T = 0$).

OVERCHARGED ($|Q| > M$)

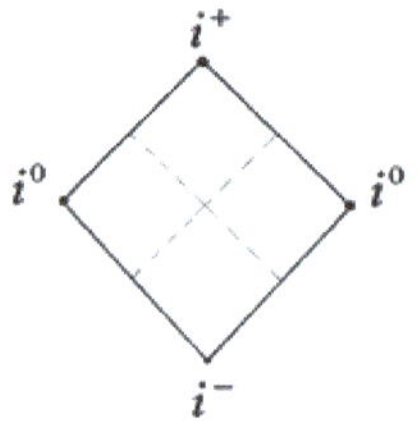

- No real horizons.
- Singularity at $r = 0$ is naked.
- Excluded by cosmic censorship conjecture.

PHYSICAL PICTURE

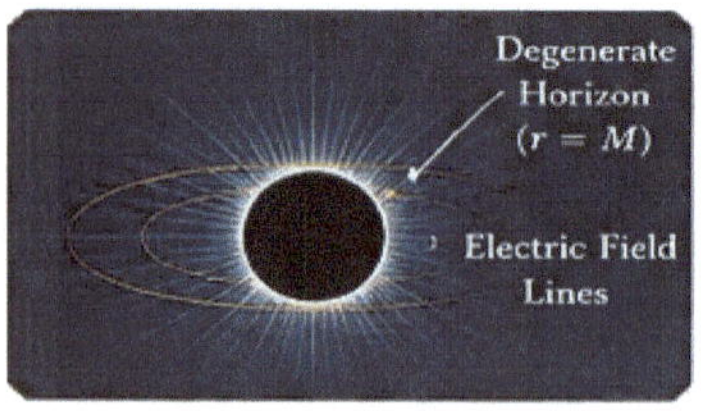

The charge creates an electric field balancing gravity exactly at the horizon. The black hole is extremal, with zero temperature and no inner region.

ELECTRIC FIELD

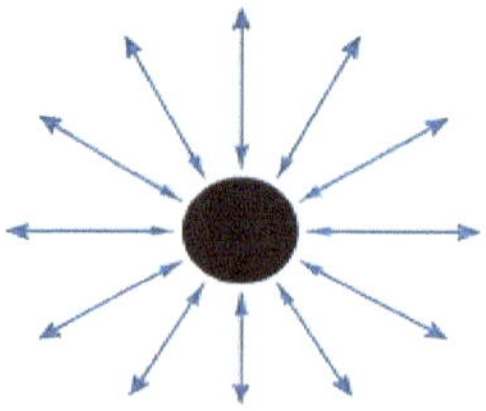

Electric field of an extremal Reissner–Nordström black hole is $E = Q/r^2$ (radially outward for $Q > 0$).

THERMODYNAMICS

Temperature:
$$T_H = \frac{\kappa}{2\pi} = 0$$

Entropy:
$$S = \frac{A}{4} = \pi M^2$$

First Law:
$$dM = \Phi_H dQ$$

Electric Potential:
$$\Phi_H = Q/r_+ = 1$$
(for $|Q| = M$)

IMPORTANT POINTS

- Extremal black holes are the boundary between black holes and naked singularities.
- They have zero temperature but non-zero entropy.
- Quantum effects may still act near the horizon.
- Play an important role in string theory and supersymmetry.

KEY TAKEAWAY

When charge equals mass ($|Q| = M$), the Reissner–Nordström black hole becomes extremal: the two horizons merge, the temperature vanishes, and the black hole sits at the threshold between stability and naked singularity.

10 Kerr Black Hole

(Rotating Black Hole)

The Kerr solution describes a rotating, uncharged black hole with mass M and angular momentum $J = aM$ (specific angular momentum $a = J/M$).

Key Features

- Axially symmetric and stationary.
- Two parameters: mass M and angular momentum $J = aM$.
- Event horizon exists for $|a| \leq M$.
- Outer and inner horizons: $r_+ \geq r_-$.
- For $|a| > M$: no horizon, naked singularity.
- Rotating horizons lead to rich phenomena: ergosphere, frame dragging, superradiance.
- Reduces to Schwarzschild when $a = 0$.

Horizon Radii

For $|a| \leq M$,

$$r_\pm = M \pm \sqrt{M^2 - a^2}$$

where $a = \dfrac{J}{M}$, $|a| \leq M$.

r_+ : outer (event) horizon

r_- : inner (Cauchy) horizon

Legend (Penrose Diagram)

- Outer Horizon ($r = r_+$)
- Inner Horizon ($r = r_-$)
- Ergosurface ($r = r_E(\theta)$)
- Null (lightlike)
- Timelike
- Spacelike

Penrose Diagram (Maximal Extension)

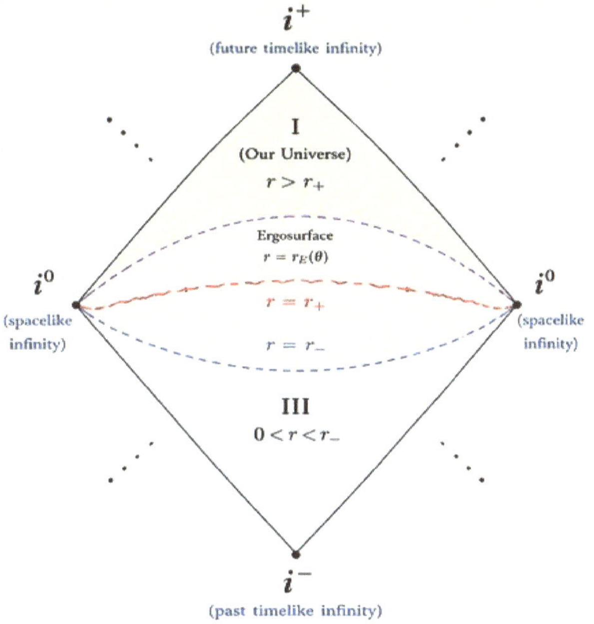

Region I is our universe.

Crossing r_+ leads inward; nothing can return to Region I.

Interpretation (Region I)

- i^+ — Future timelike infinity: where future-directed timelike paths end.
- i^- — Past timelike infinity: where past-directed timelike paths begin.
- i^0 — Spacelike infinity: points at spatial infinity at one instant.
- Outer horizon ($r = r_+$): one-way boundary; no return once crossed.
- Inner horizon ($r = r_-$): Cauchy horizon; unstable to perturbations.
- Ergosurface ($r = r_E(\theta)$): outside the horizon where no static observers exist.
- Frame dragging is so strong inside ergosphere that all objects are compelled to corotate.

Exists Horizon if and only if

$$|a| \leq M$$

- $|a| < M$: non-extremal (two horizons)
- $|a| = M$: extremal (horizons coincide)
- $|a| > M$: no horizon (naked singularity)

Ergosphere and Frame Dragging

Ergosurface:

$$r_E(\theta) = M + \sqrt{M^2 - a^2 \cos^2\theta}$$

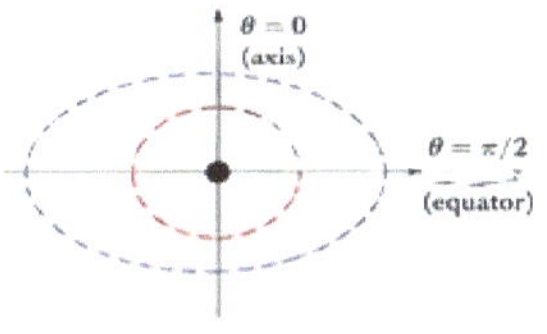

- - - Ergosurface (r_E)
— Event Horizon (r_+)

- Outside $r_E(\theta)$: static observers possible.
- Between $r_+ < r < r_E$: no static observers (must co-rotate).
- Inside r_+: cannot return to outside.

Frame dragging: spacetime itself is dragged around the black hole.

Horizon Properties

Property	Value (at r_+)
Angular Velocity	$\Omega_H = \dfrac{a}{2Mr_+}$
Surface Gravity	$\kappa = \dfrac{r_+ - r_-}{2(r_+^2 + a^2)}$
Area	$A = 4\pi(r_+^2 + a^2)$
Irreducible Mass	$M_{\rm irr} = \sqrt{\dfrac{A}{16\pi}}$
Horizon Generator	$\chi = \partial_t + \Omega_H \partial_\phi$

Temperature and Entropy

Hawking Temperature

$$T_H = \frac{\kappa}{2\pi} = \frac{r_+ - r_-}{4\pi(r_+^2 + a^2)}$$

Bekenstein–Hawking Entropy

$$S = \frac{A}{4} = \pi(r_+^2 + a^2)$$

- $T_H = 0$ as $|a| \to M$ (extremal limit).
- Entropy depends only on horizon area.
- Rotation increases the horizon area for fixed mass.

Physical Picture

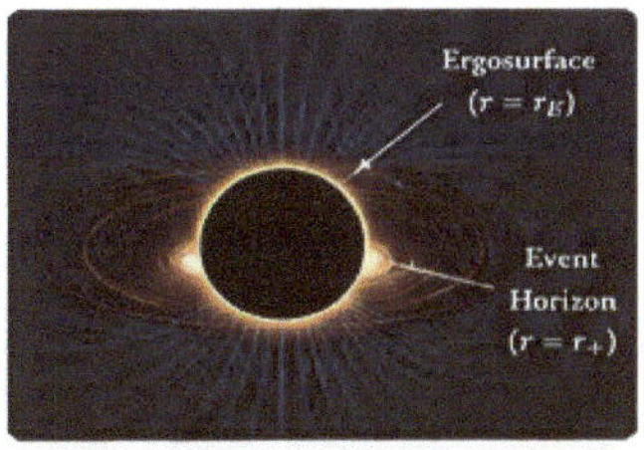

Inside the ergosphere, spacetime is dragged around the black hole. Energy extraction mechanisms become possible.

Rotational Energy Extraction

Penrose Process (inside ergosphere)

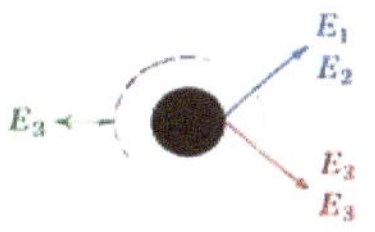

- A particle splits inside the ergosphere.
- One fragment can have negative energy ($E_2 < 0$).
- Other fragments can escape with $E_1 + E_3 > E_{\rm in}$.

Superradiance (Waves)

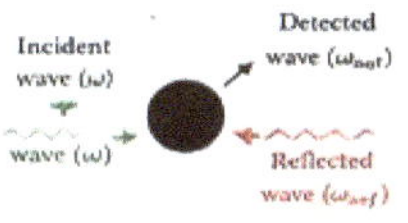

- For $\omega < m\Omega_H$, reflected wave is amplified.
- Energy and angular momentum are extracted.
- Classical wave version of the Penrose process.

Electric Field

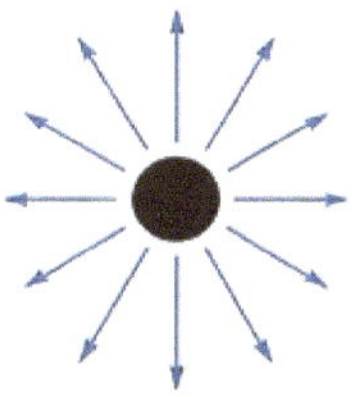

Electric field of an extremal Reissner–Nordström black hole is $E = Q/r^2$ (radially outward for $Q > 0$).

Important Points

- Kerr black holes are the most realistic astrophysical black holes.
- Ergosphere enables energy extraction (Penrose, BZ, etc.).
- Cosmic censorship conjecture suggests $|a| \leq M$ in nature.
- Kerr geometry is fundamental in accretion disks, jets, and gravitational waves.

Key Takeaway

Rotation enriches black hole physics: horizons, ergosphere, and frame dragging lead to powerful new phenomena—energy extraction, superradiance, and deep connections between stability and cosmic censorship.

11 KERR BLACK HOLE

(MAXIMAL EXTENSION)

The maximal extension of the Kerr solution reveals an infinite, repeating lattice of regions connected through horizons and ring singularities.

KEY FEATURES

- Infinite sequence of regions labeled by integers (n, m).
- Horizons at $r = r_+$ separate regions; $r = r_+$ is the event horizon.
- Ring singularity at $r = 0$, $\theta = \pi/2$ in all regions.
- The structure is periodic in time and azimuthal angle.
- Every region has an asymptotically flat universe (Region I).

HORIZON RADII

For Kerr black hole,

$$r_\pm = M \pm \sqrt{M^2 - a^2}$$

where $a = J/M, \ |a| \le M$.

r_+ : outer (event) horizon

r_- : inner (Cauchy) horizon

LEGEND (PENROSE DIAGRAM)

- Outer horizon $(r = r_+)$
- Inner horizon $(r = r_-)$
- Ring singularity $(r = 0, \ \theta = \pi/2)$
- Timelike boundary
- Null (lightlike) boundary
- Spacelike boundary

PENROSE DIAGRAM (MAXIMAL EXTENSION)

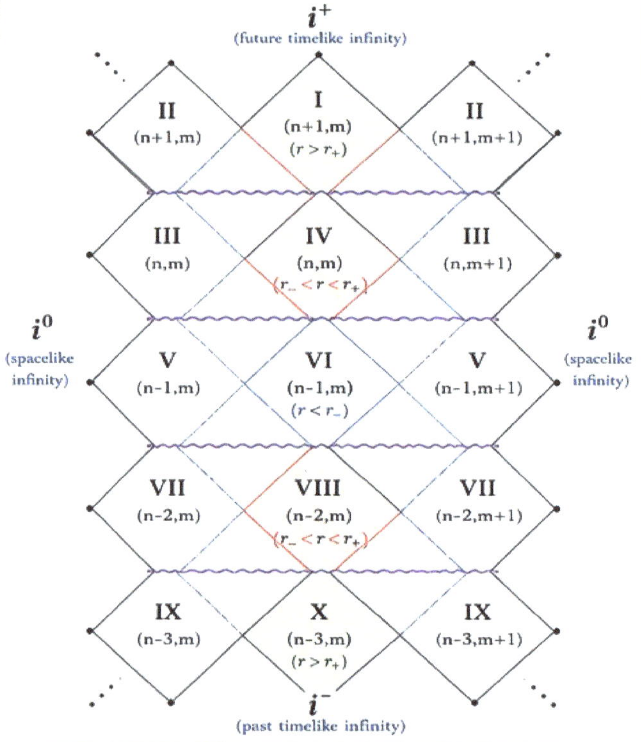

Each column contains one asymptotically flat universe (Region I). Regions repeat infinitely in both time and space directions.

INTERPRETATION

- i^+ Future timelike infinity: where future-directed timelike paths end.
- i^- Past timelike infinity: where past-directed timelike paths begin.
- i^0 Spacelike infinity: points at spatial infinity in each universe.
- Outer horizon $(r = r_+)$: one-way boundary; cannot return once crossed inward.
- Inner horizon $(r = r_-)$: Cauchy horizon; signals cannot escape its interior.
- Ring singularity $(r = 0, \ \theta = \pi/2)$: a closed ring in every region; curvature diverges.
- Maximal extension is regular everywhere except at the ring singularity.

REGION MEANING (n, m)

- n : counts progress in (advanced) time.
- m : counts progress in azimuthal angle ϕ.
- Even/odd n alternate between asymptotically flat and non-flat regions.
- The spacetime is symmetric under $n \to n \pm 2, \ m \to m \pm 2$.

TYPES OF REGIONS

Region I (Our Universe)

$r > r_+$

- Outside the event horizon.
- Asymptotically flat.
- Stationary observers possible far from the black hole.

Region IV / VIII (Ergospheric Regions)

$r_- < r < r_+$

- Between horizons.
- No static observers.
- Frame dragging is so strong that all objects co-rotate.

Region II / III / V / VII / IX ... (Other Exterior-like Regions)

$r > r_+$

- Asymptotically flat.
- Represent other universes in the extended spacetime.

Region VI / X / ... (Inner Regions)

$r < r_-$

- Inside the Cauchy horizon.
- Time and radial roles interchange.
- Curvature finite except at ring singularity.

Ring Singularity

$r = 0, \ \theta = \pi/2$

- Present in every region.
- A closed ring.
- True physical singularity.

PHYSICAL PICTURE

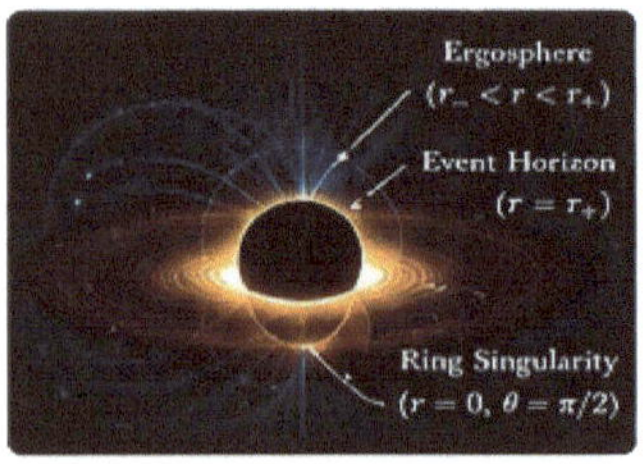

Frame dragging causes spacetime itself to rotate around the black hole. Energy extraction is possible.

FRAME DRAGGING

Angular Velocity of Dragging

$$\omega(r, \theta) = \frac{2aMr}{\Sigma} \, g_{\phi\phi}$$

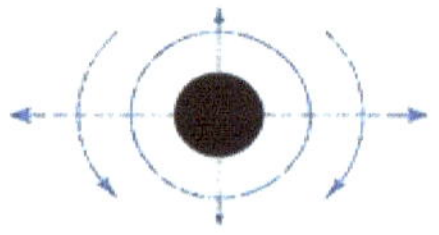

- $\omega > 0$ outside the horizon.
- In the ergosphere, no static observers exist.
- Objects are compelled to co-rotate.

IMPORTANT QUANTITIES

Quantity	Expression
Horizon Radii	$r_\pm = M \pm \sqrt{M^2 - a^2}$
Angular Velocity (Horizon)	$\Omega_H = \dfrac{a}{2Mr_+}$
Horizon Area	$A = 4\pi(r_+^2 + a^2)$
Surface Gravity	$\kappa = \dfrac{r_+ - r_-}{2(r_+^2 + a^2)}$
Hawking Temp.	$T_H = \dfrac{\kappa}{2\pi}$
Entropy	$S = \dfrac{A}{4} = \pi(r_+^2 + a^2)$

WHY IT MATTERS

- The Kerr maximal extension shows the full causal structure of rotating black holes.
- Understanding horizons and the ergosphere is key to energy extraction (Penrose, BZ, etc.).
- Reveals deep symmetries and connections between distant regions and universes.
- Essential for quantum gravity, information flow, and cosmic censorship studies.

KEY TAKEAWAY

The Kerr maximal extension unveils a rich and infinite spacetime: horizons partition regions, the ring singularity lies at the core, and rotation creates an ergosphere where spacetime itself is dragged—connecting universes through a beautiful cosmic lattice.

12 KERR–NEWMAN BLACK HOLE

(ROTATING CHARGED BLACK HOLE)

The Kerr–Newman solution describes a black hole with mass M, charge Q, and angular momentum $J = aM$. It generalizes Kerr ($Q = 0$) and Reissner–Nordström ($a = 0$).

KEY FEATURES

- Most general stationary, axisymmetric, asymptotically flat black hole of GR.
- Characterized by M, Q, and $a = J/M$, $|a| \le M$.
- Two horizons when $M^2 > a^2 + Q^2$ (subextremal).
- One degenerate horizon when $M^2 = a^2 + Q^2$ (extremal).
- No horizon (naked singularity) when $M^2 < a^2 + Q^2$ (overcharged or overspun).
- Rich causal structure with ergosphere and frame dragging.

HORIZON RADII

The outer (event) and inner (Cauchy) horizon radii are

$$r_\pm = M \pm \sqrt{M^2 - a^2 - Q^2}$$

Existence of horizons requires $M^2 \ge a^2 + Q^2$.

r_+ : event horizon

r_- : Cauchy horizon

LEGEND (PENROSE DIAGRAM)

- Outer horizon ($r = r_+$)
- Inner horizon ($r = r_-$)
- Ergosurface ($r = r_E(\theta)$)
- Null (lightlike) boundary
- Timelike boundary
- Spacelike boundary
- Singularity ($r = 0, \theta = \pi/2$)

PENROSE DIAGRAM (SUBEXTREMAL: $M^2 > a^2 + Q^2$)

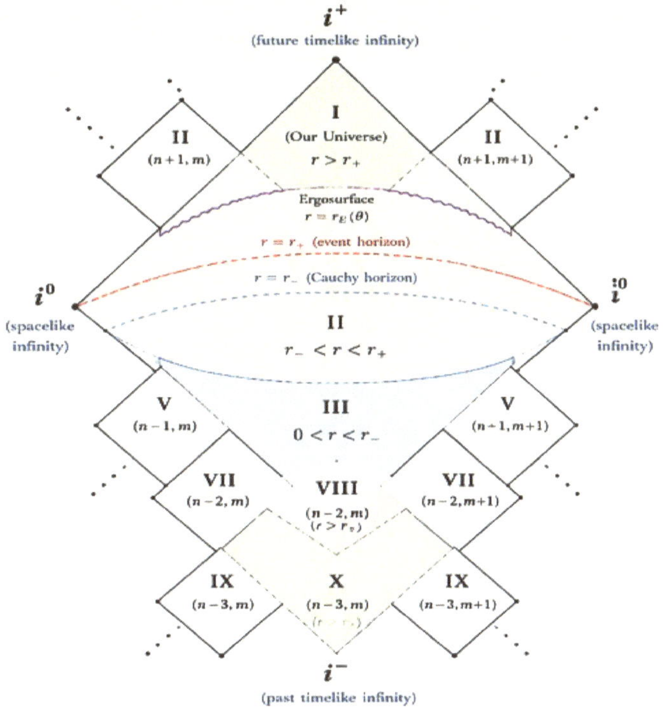

Region I is our universe. Crossing r_+ leads inward. Crossing r_- leads to another region. Crossing $r = 0$ leads to $r < 0$ and returns outward. The ring singularity lies behind r_- in the equatorial plane.

INTERPRETATION (REGION)

- i^+ — Future timelike infinity: where future-directed timelike paths end.
- i^- — Past timelike infinity: where past-directed timelike paths begin.
- i^0 — Spacelike infinity: points at spatial infinity in all directions.
- Outer horizon ($r = r_+$): one-way boundary; no return once crossed.
- Inner horizon ($r = r_-$): Cauchy horizon; unstable and leads to infinities.
- Ergosurface ($r = r_E(\theta)$): outside r_+ where no static observers exist.
- Singularity ($r = 0$, $\theta = \pi/2$): a ring of curvature singularity.

CONDITIONS AND REGIMES

Condition	Regime
$M^2 > a^2 + Q^2$	Subextremal (two horizons)
$M^2 = a^2 + Q^2$	Extremal (degenerate horizon)
$M^2 < a^2 + Q^2$	Overcharged / Overspun (no horizon)

SPECIAL CASES

KERR ($Q = 0$)

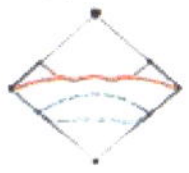

- Two horizons $r_\pm = M \pm \sqrt{M^2 - a^2}$.
- Reduces to Schwarzschild when $a = 0$.

REISSNER–NORDSTRÖM ($a = 0$)

- No ergosphere.
- Two horizons $r_\pm = M \pm \sqrt{M^2 - Q^2}$.
- Reduces to Schwarzschild when $Q = 0$.

SCHWARZSCHILD ($a = 0$, $Q = 0$)

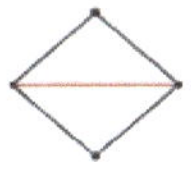

- Single horizon $r_+ = 2M$.
- Spherically symmetric.

ROLE OF PARAMETERS

Parameter	Effect	Physical Meaning
M	Increases $r_\pm$ and area	Total mass–energy
Q	Increases r_+; reduces gravity	Electric charge
a	Increases frame dragging; creates ergosphere	Angular momentum
$r_E(\theta)$	Ergosurface radius depends on θ	Latitude-dependent dragging region

PHYSICAL PICTURE

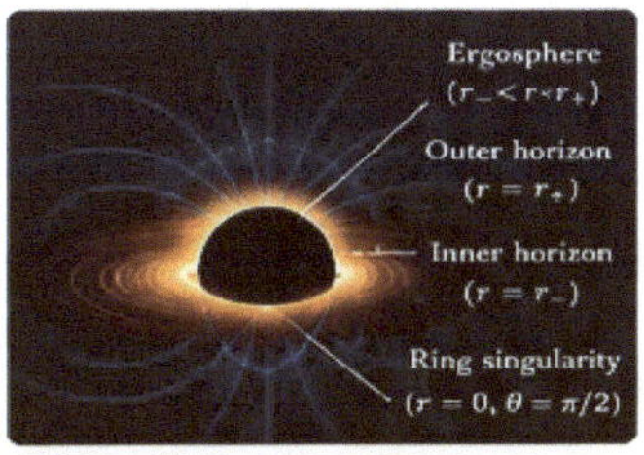

Frame dragging causes spacetime itself to rotate around the black hole. Energy extraction is possible.

FRAME DRAGGING

Angular velocity of dragging (ZAMO frame):

$$\omega(r, \theta) = -\frac{2aMr}{\Sigma}$$

$$\Sigma = r^2 + a^2 \cos^2\theta$$

- $\omega > 0$ outside the horizon.
- In the ergosphere, no static observers exist.
- Objects are compelled to co-rotate.

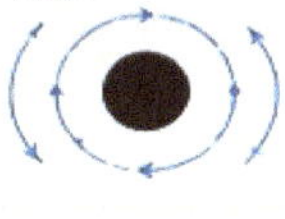

IMPORTANT QUANTITIES

Quantity	Expression
Horizon Radii	$r_\pm = M \pm \sqrt{M^2 - a^2 - Q^2}$
Angular Velocity (Horizon)	$\Omega_H = \dfrac{a}{2Mr_+ + Q^2}$
Horizon Area	$A = 4\pi(r_+^2 + a^2)$
Surface Gravity	$\kappa = \dfrac{r_+ - r_-}{2(r_+^2 + a^2)}$
Hawking Temp.	$T_H = \dfrac{\kappa}{2\pi}$
Electric Potential	$\Phi_H = \dfrac{Qr_+}{r_+^2 + a^2}$
Entropy	$S = \dfrac{A}{4} = \pi(r_+^2 + a^2)$

KEY INSIGHTS

- Kerr–Newman black holes unify mass, charge, and rotation.
- Ergosphere enables energy extraction (Penrose process, BZ mechanism).
- Inner horizon is unstable (mass inflation); quantum effects may regularize it.
- Extremal limit has zero temperature but finite entropy.
- Real astrophysical black holes are expected to be near Kerr.

KEY TAKEAWAY

The Kerr–Newman maximal extension reveals a richly structured spacetime. Horizons, ergosphere, and ring singularity arise from the interplay of mass, charge, and rotation—unlocking profound physics and cosmic phenomena.

13 De Sitter Spacetime

(Cosmological Constant $\Lambda > 0$)

De Sitter spacetime is the maximally symmetric solution of Einstein's equations with a positive cosmological constant $\Lambda > 0$ and no matter.

Key Features

- Maximally symmetric, vacuum solution with $\Lambda > 0$.
- No mass, no charge, no rotation.
- Positive cosmological constant acts like uniform vacuum energy.
- No singularity: spacetime is geodesically complete.
- All observers see a cosmological horizon.
- Enormously important in cosmology (inflation, dark energy).

Cosmological Horizon

In static coordinates, the cosmological horizon radius is

$$r_c = \sqrt{\frac{3}{\Lambda}}$$

Beyond r_c, static observers cannot receive signals.

Legend (Penrose Diagram)

- Cosmological horizon ($r = r_c$)
- Null (lightlike) boundary
- Timelike boundary
- Spacelike boundary
- Worldline of static observer
- Constant cosmic time slices

Penrose Diagram (Global)

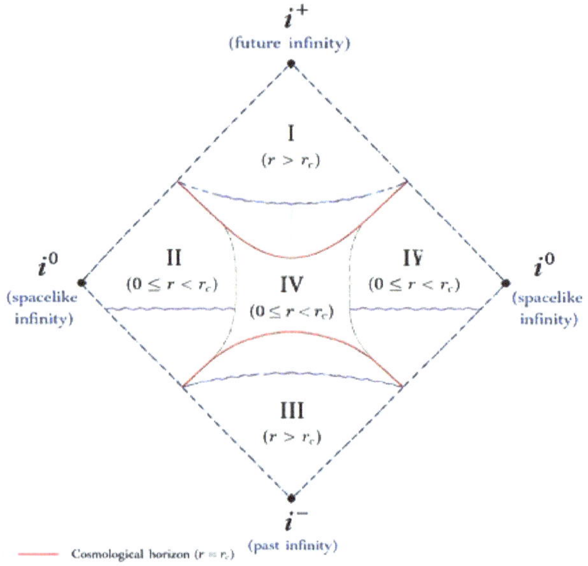

- Cosmological horizon ($r \approx r_c$)
- Null (lightlike) boundary
- Timelike boundary
- Spacelike boundary
- Worldline of static observer
- Constant cosmic time slices

- Regions I and III: beyond the cosmological horizon ($r > r_c$).
- Regions II and IV: static regions ($0 \le r < r_c$) where observers can remain at rest.
- No singularities anywhere; spacetime is smooth.

Interpretation

- i^+ Future infinity: all timelike paths end here.
- i^- Past infinity: all timelike paths begin here.
- i^0 Spacelike infinity: spatial infinity at one instant.
- Cosmological horizon ($r = r_c$): boundary beyond which events cannot affect us.
- No black hole: nothing trapped inside.
- Observers in II or IV see a horizon and Gibbons–Hawking temperature.
- Spacetime is eternal, maximally symmetric, and geodesically complete.

Metric (Static Patch)

In static coordinates ($0 \le r < r_c$),

$$ds^2 = -\left(1 - \frac{r^2}{r_c^2}\right) dt^2 + \left(1 - \frac{r^2}{r_c^2}\right)^{-1} dr^2 + r^2 d\Omega_2^2$$

with $r_c = \sqrt{\frac{3}{\Lambda}}$ and $d\Omega_2^2 = d\theta^2 + \sin^2\theta d\phi^2$.
Valid only for $0 \le r < r_c$.

Properties of De Sitter Spacetime

Property	Expression / Value	Meaning
Einstein Equation	$R_{\mu\nu} = \Lambda g_{\mu\nu}$	Vacuum solution with positive Λ.
Ricci Scalar	$R = 4\Lambda$	Constant positive curvature.
Weyl Tensor	$C_{\mu\nu\rho\sigma} = 0$	Conformally flat spacetime.
Curvature Radius	$r_c = \sqrt{\frac{3}{\Lambda}}$	Sets the length scale of curvature.
Volume	$V_4 = \frac{8\pi^2}{3} r_c^4$ (4-sphere)	Finite total four-volume.

Gibbons–Hawking Temperature

Observers inside the static region ($r < r_c$) perceive a thermal bath with temperature

$$T_{\rm dS} = \frac{H}{2\pi} = \frac{1}{2\pi r_c} = \frac{1}{2\pi}\sqrt{\frac{\Lambda}{3}}$$

where $H = \sqrt{\frac{\Lambda}{3}}$ is the Hubble constant in de Sitter space.

Embedding in 5D Minkowski Space

De Sitter space can be embedded as a hyperboloid in 5D Minkowski space with metric

$$-X_0^2 + X_1^2 + X_2^2 + X_3^2 + X_4^2 = r_c^2$$

in 5D Minkowski space with metric

$$ds^2 = -dX_0^2 + \sum_{i=1}^{4} dX_i^2.$$

This manifold has maximal symmetry SO(1, 4).

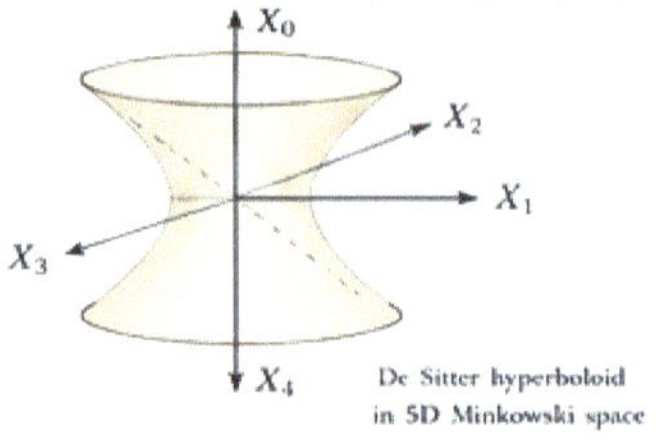

Volume of Static Patch

The 4-volume of the static patch ($0 \le r < r_c$) is finite:

$$V_{\rm patch} = \frac{\pi^2}{2} r_c^4 = \frac{9\pi^2}{2\Lambda^2}$$

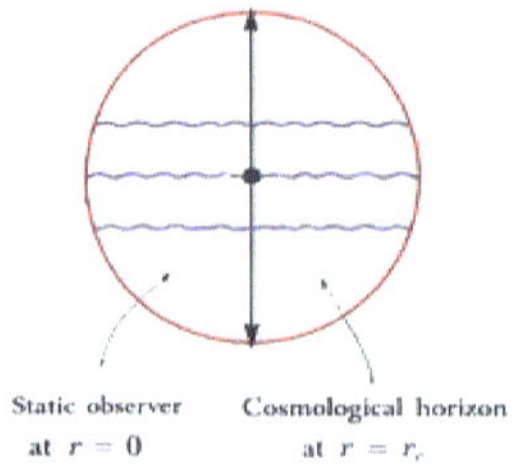

Conformal Diagram (Square)

De Sitter space is conformal to a square with identified edges.

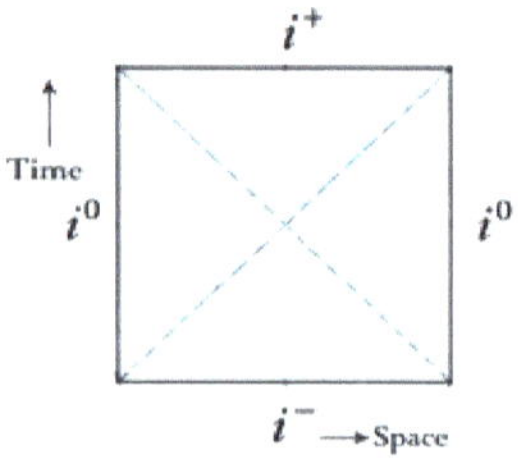

Opposite edges are identified. All observers share the same causal structure.

Importance in Cosmology

- Models the inflationary universe in the early cosmos.
- Provides a simple model for dark energy–driven acceleration.
- Has a natural temperature, linking gravity, thermodynamics, and quantum theory.
- Serves as a testing ground for quantum gravity and holography.
- Basis of de Sitter/CFT correspondence and quantum information studies.

Key Takeaway

De Sitter spacetime is the simplest solution with $\Lambda > 0$: no singularities, a cosmological horizon, thermal radiation, and maximal symmetry. It plays a central role in modern cosmology and quantum gravity.

14

ANTI–DE SITTER SPACETIME

($\lambda < 0$)

Anti–de Sitter (AdS) spacetime is a maximally symmetric solution of Einstein's equations with a negative cosmological constant $\Lambda < 0$.

KEY FEATURES

- Maximally symmetric solution with $\Lambda < 0$.
- No mass, no charge, no rotation.
- Constant negative curvature.
- No event horizon.
- Timelike infinity: boundaries are timelike.
- Rich boundary structure at spatial infinity.
- Central role in the AdS/CFT correspondence.

CURVATURE RADIUS

The curvature radius (AdS radius) is given by

$$\ell = \sqrt{\frac{-3}{\Lambda}}$$

where $\Lambda < 0$.
Smaller ℓ means stronger (curvature magnitude).

LEGEND (PENROSE DIAGRAM)

- Timelike boundary (conformal)
- Null (lightlike) boundary
- Timelike surface
- Spacelike surface
- Worldline of static observer
- Constant time slices

PENROSE DIAGRAM (GLOBAL ADS)

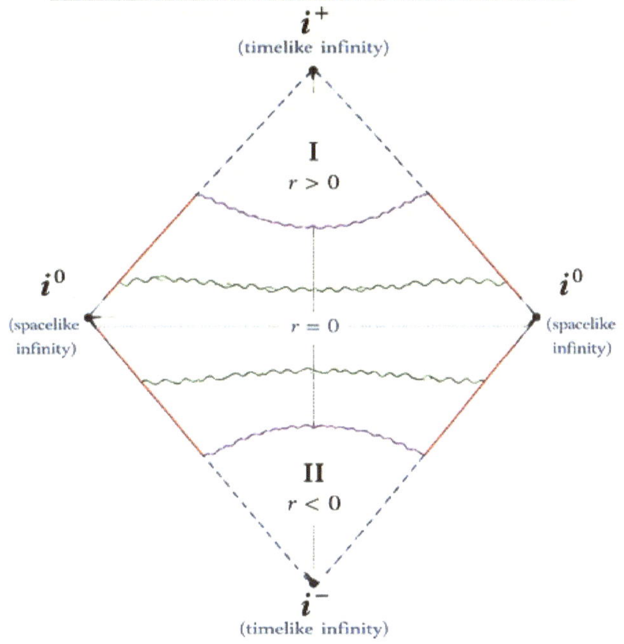

- Timelike boundary at spatial infinity (conformal boundary).
- All observers, signals, and particles remain within the spacetime.
- No horizons: entire spacetime is visible in principle.
- Negative curvature gives "saddle–like" geometry.

INTERPRETATION

- i^+ Future timelike infinity: all future-directed timelike paths end here.
- i^- Past timelike infinity: all past-directed timelike paths start here.
- i^0 Spacelike infinity: boundary at spatial infinity is timelike.
- Timelike boundary ($r = \ell$): conformal boundary where fields can live (AdS/CFT).
- No horizons or singularities in pure AdS spacetime.
- ∞ Spacetime is geodesically complete and globally hyperbolic.

METRIC (GLOBAL COORDINATES)

The global AdS metric in $(d + 1)$ dimensions:

$$ds^2 = -\left(1 + \frac{r^2}{\ell^2}\right) dt^2 + \left(1 + \frac{r^2}{\ell^2}\right)^{-1} dr^2 + r^2 d\Omega_{d-1}^2$$

where $\ell = \sqrt{\frac{-d(d-1)}{2\Lambda}}$, $\Lambda < 0$,

and $d\Omega_{d-1}^2$ is the metric on S^{d-1}.

GEOMETRIC PROPERTIES

Property	Expression / Value	Meaning
Einstein Equation	$R_{\mu\nu} = \Lambda g_{\mu\nu}$	Vacuum with negative Λ.
Ricci Scalar	$R = (d+1)\Lambda$	Constant negative curvature.
Weyl Tensor	$C_{\mu\nu\rho\sigma} = 0$	Conformally flat spacetime.
Curvature Radius	$\ell = \sqrt{\frac{-d(d-1)}{2\Lambda}}$	Sets curvature scale.
Volume (t = const.)	$V \propto \int_0^\infty \frac{r^{d-1}}{\sqrt{1 + r^2/\ell^2}}\, dr$	Infinite (but can be regularized).

THERMAL PROPERTIES

- Pure AdS spacetime has no horizon → no intrinsic Hawking temperature.
- It is in its ground (vacuum) state.
- When a black hole exists in AdS (Schwarzschild–AdS, Kerr–AdS), it can have a Hawking temperature.

For Schwarzschild–AdS black hole:

$$T_H = \frac{1}{4\pi r_+}\left((d-2) + d\,\frac{r_+^2}{\ell^2}\right)$$

Large black holes in AdS have positive specific heat and are thermodynamically stable.

EMBEDDING IN HIGHER DIMENSION

AdS_{d+1} can be embedded in $\mathbb{R}^{2,d}$ as the hyperboloid:

$$-X_0^2 - X_d^2 + X_1^2 + \cdots + X_{d-1}^2 = \ell^2$$

with metric $ds^2 = -dX_0^2 + \sum_{i=1}^{d} dX_i^2$.

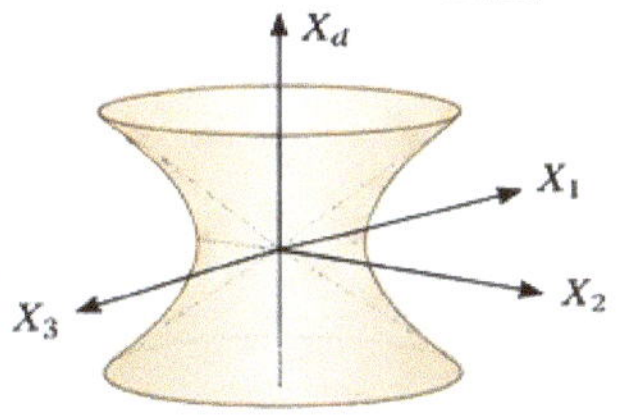

A two-sheeted hyperboloid.
AdS is one of its sheets.

SPATIAL GEOMETRY (t = const.)

Spatial slices are hyperbolic spaces H^d with constant negative curvature $-1/\ell^2$.

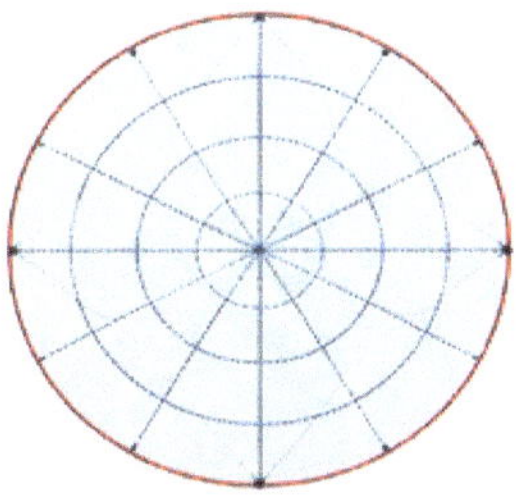

Poincaré disk model (example: H^2).
Boundary (red circle) is at infinity.

CONFORMAL BOUNDARY

The timelike boundary at $r = \ell$ is conformally $\mathbb{R} \times S^{d-1}$.

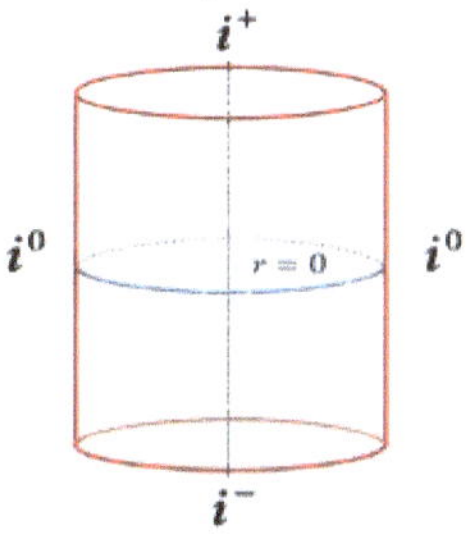

Boundary is timelike and acts as a "box wall" of AdS.

AdS/CFT CORRESPONDENCE

A fundamental duality relates gravity in AdS_{d+1} to a conformal field theory (CFT_d) on the boundary.

Gravity in AdS_{d+1} (bulk) —Dual→ CFT_d on ∂ AdS (boundary)

- Bulk fields ↔ boundary operators
- Partition functions are equal
- Strong/weak coupling duality
- Powerful tool in quantum gravity, holography, and condensed matter.

KEY TAKEAWAY

Anti–de Sitter spacetime has no horizons and is geodesically complete.
Its timelike boundary at infinity makes it unique and crucial for holography.
It provides a natural laboratory for quantum gravity and strongly coupled field theories.

15 Friedmann–Lemaître–Robertson–Walker (FLRW) Spacetime

(Homogeneous and Isotropic Universes)

FLRW spacetimes describe a universe that is homogeneous and isotropic on large scales. They are the foundation of modern cosmology.

Key Features

- Homogeneous: the same at every location.
- Isotropic: the same in every direction.
- Described by a scale factor $a(t)$.
- Curvature index: $k = +1, 0, -1$ (closed, flat, open).
- Solutions depend on matter content and Λ.
- Forms the basis of the Standard Cosmological Model (ΛCDM).

Curvature Parameter k

Determines spatial geometry of constant-time slices:

$k = +1$: closed (spherical)
$k = 0$: flat (Euclidean)
$k = -1$: open (hyperbolic)

Penrose Diagram (Typical FLRW)

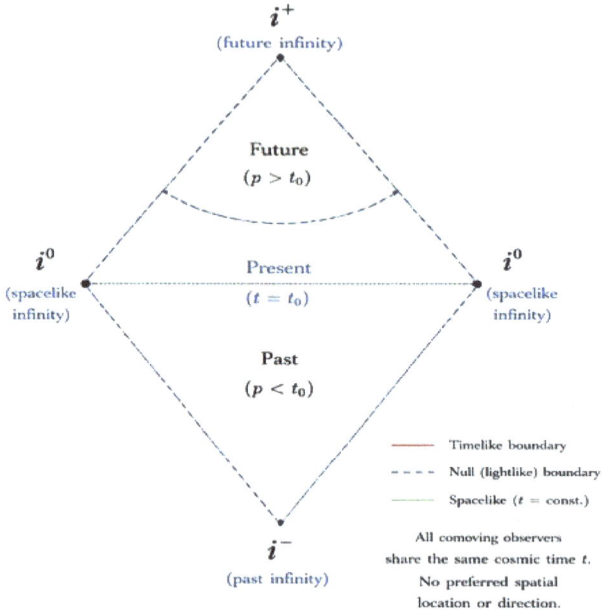

All comoving observers share the same cosmic time t. No preferred spatial location or direction.

Interpretation

- i^+ Future infinity: the universe evolves forward in time.
- i^- Past infinity: (or a beginning at finite time).
- i^0 Spacelike infinity: very large spatial distances.
- Time slices (t = const.) are space-like hypersurfaces.
- No event horizon in the ideal FLRW universe.
- The universe is the same everywhere and in every direction.

Comoving Observer

Observers at rest in these coordinates see the universe expand but do not move relative to the cosmic fluid.

Their 4-velocity is $u^\mu = (1, 0, 0, 0)$ in comoving coordinates.

FLRW Metric

The line element of FLRW spacetime is

$$ds^2 = -c^2 dt^2 + a^2(t)\left[\frac{dr^2}{1 - kr^2} + r^2\,(d\theta^2 + \sin^2\theta\, d\phi^2)\right]$$

where

- t = cosmic time
- $a(t)$ = scale factor
- (r, θ, ϕ) = comoving spherical coordinates
- $k = +1, 0, -1$ = curvature index

Scale Factor $a(t)$

- Measures the relative expansion of the universe.
- $a(t_0) = 1$ today (by convention).
- Physical distance between comoving points:

$$D(t) = a(t)\,\chi$$

where χ is the comoving distance.

Redshift

Light emitted at t_e and observed at t_0 has

$$1 + z = \frac{a(t_0)}{a(t_e)}$$

Cosmic expansion stretches wavelengths.

Spatial Geometry (t = const.)

$k = +1$ (Closed)	$k = 0$ (Flat)	$k = -1$ (Open)
3-sphere S^3	Euclidean $\mathbb{R}^3$	Hyperbolic H^3

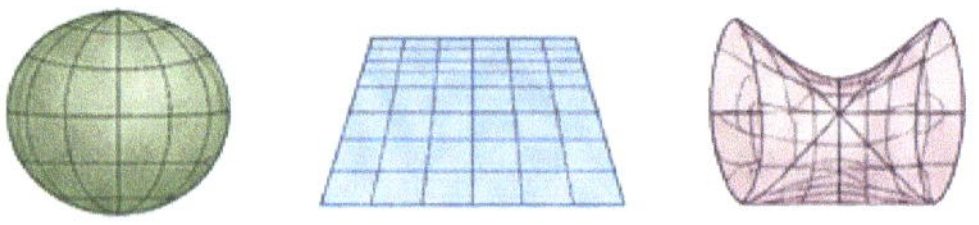

$0 \le r < 1$	$0 \le r < \infty$	$0 \le r < \infty$
Positive curvature.	Zero curvature.	Negative curvature.

Dynamics: Friedmann Equations

Derived from Einstein's equations

(1) First Friedmann equation

$$\left(\frac{\dot a}{a}\right)^2 + \frac{kc^2}{a^2} = \frac{8\pi G}{3}\rho + \frac{\Lambda c^2}{3}$$

(2) Acceleration equation

$$\frac{\ddot a}{a} = -\frac{4\pi G}{3}\left(\rho + \frac{3p}{c^2}\right) + \frac{\Lambda c^2}{3}$$

(3) Continuity equation

$$\dot\rho + 3\frac{\dot a}{a}\left(\rho + \frac{p}{c^2}\right) = 0$$

$$H \equiv \frac{\dot a}{a} = \text{Hubble parameter.}$$

Density Parameters

Define dimensionless density parameters.

$$\Omega_m = \frac{8\pi G\rho_m}{3H^2} \qquad \Omega_r = \frac{8\pi G\rho_r}{3H^2}$$

$$\Omega_\Lambda = \frac{\Lambda c^2}{3H^2} \qquad \Omega_k = -\frac{kc^2}{a^2H^2}$$

They satisfy the closure relation

$$\Omega_m + \Omega_r + \Omega_\Lambda + \Omega_k = 1$$

Cosmic History (Qualitative)

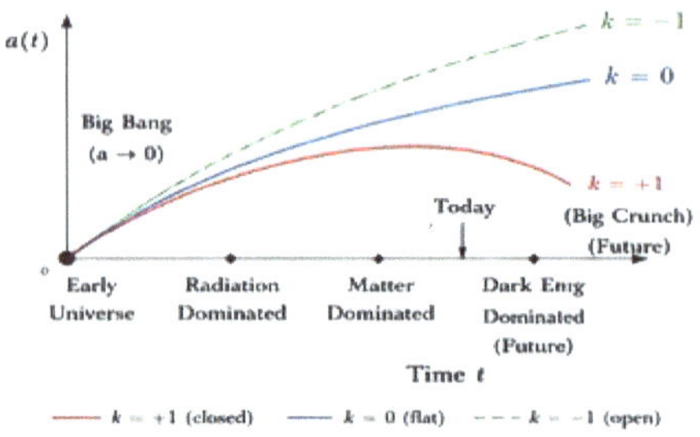

Special Cases

Einstein–de Sitter universe:
$\Omega_m = 1,\ \Omega_\Lambda = 0,\ k = 0.$
Scale factor
$a(t) \propto t^{2/3},\ \ \Omega_m(t) = t^0.$

Milne universe: $k = -1,\ p = 0,\ \Lambda = 0.$
$a(t) \propto t$ (coasting).

Inflationary phase: $a(t) \propto e^{Ht}$ with nearly constant H.

Our universe (today):
$\Omega_m \approx 0.3,\ \ \Omega_\Lambda \approx 0.7,\ \ \Omega_k \approx 0.$

Why FLRW is Important

- Matches cosmic microwave background (CMB) isotropy.
- Explains redshift–distance relation (Hubble law).
- Provides evolution of the universe from the Big Bang.
- Foundation for structure formation, cosmic inflation, and dark energy studies.
- Simplicity with rich physics: a cornerstone of modern cosmology.

Key Takeaway

FLRW spacetimes capture the large-scale symmetry of our universe. The scale factor $a(t)$ encodes cosmic history, while Friedmann equations connect the geometry of spacetime to its energy content.

16 FLRW Solutions and Curved Spacetime

Geometry, Expansion, and the Fate of the Universe

The curvature index k and matter—energy content determine the dynamics of the scale factor $a(t)$, the age of the universe, and its ultimate destiny.

Curvature Geometry (t = const.)

$k = +1$ (Closed)

3-sphere S^3

- Positive curvature
- Finite, unbounded
- Volume finite

$k = 0$ (Flat)

Euclidean $\mathbb{R}^3$

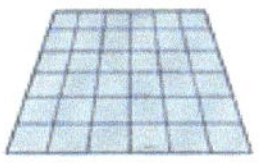

- Zero curvature
- Infinite, unbounded
- Volume infinite

$k = -1$ (Open)

Hyperbolic H^3

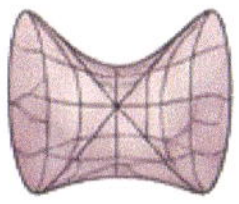

- Negative curvature
- Infinite, unbounded
- Volume infinite

Penrose Diagrams (vs. r)

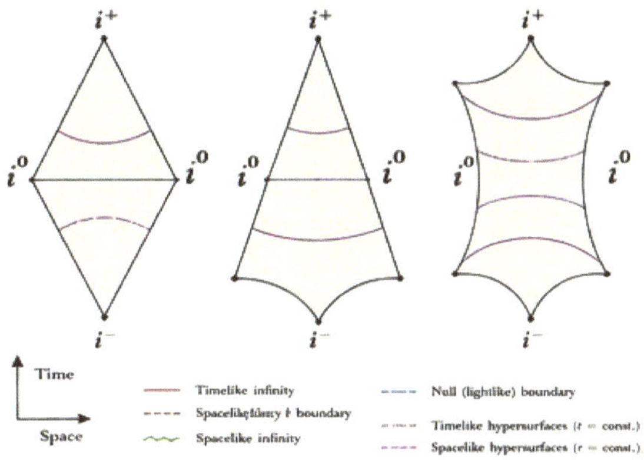

Interpretation

- i^+ Future infinity: the universe may keep expanding forever or end in a crunch.
- i^- Past infinity (or beginning): Big Bang for standard cosmology.
- i^0 Spacelike infinity: infinitely far away at a given time.
- Curved spacetime affects light paths and causal structure.
- All comoving observers share the same cosmic time t.
- Geometry (k) + content determine evolution).

Friedmann Equation

The first Friedmann equation governs the expansion of the universe:

$$\left(\frac{\dot{a}}{a}\right)^2 = H^2 = \frac{8\pi G}{3}\rho - \frac{kc^2}{a^2} + \frac{\Lambda c^2}{3}$$

where

- $a(t)$: scale factor
- $H = \frac{\dot{a}}{a}$: Hubble parameter
- ρ : total energy density
- $k = +1, 0, -1$: curvature index
- Λ : cosmological constant

Acceleration Equation

The second Friedmann equation gives acceleration:

$$\frac{\ddot{a}}{a} = -\frac{4\pi G}{3}\left(\rho + \frac{3p}{c^2}\right) + \frac{\Lambda c^2}{3}$$

where

- p : pressure
- $\ddot{a} = \frac{d^2a}{dt^2}$: second derivative of $a(t)$
- Acceleration occurs when $\ddot{a} > 0$, i.e. for sufficiently low $\rho + 3p/c^2$ or large Λ.

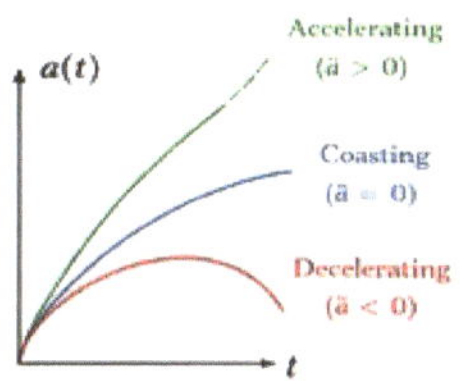

Density Parameter

Define the critical density

$$\rho_c = \frac{3H^2}{8\pi G}$$

The density parameter is

$$\Omega = \frac{\rho}{\rho_c}$$

- $\Omega > 1 \rightarrow k = +1$ (closed)
- $\Omega = 1 \rightarrow k = 0$ (flat)
- $\Omega < 1 \rightarrow k = -1$ (open)

Dynamics and Fate of the Universe (with $\Lambda = 0$)

Curvature (k)	Dominant Content	Condition (Ω_0)	Behavior of $a(t)$	Age of Universe (approx.)	Fate of Universe	Penrose Diagram
+1 (Closed)	Matter	$\Omega_0 > 1$	Expands, reaches a_{max}, then recollapses.	Finite	Big Crunch in finite time.	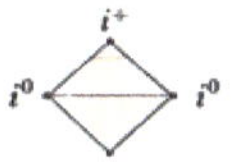
0 (Flat)	Matter	$\Omega_0 = 1$	Expands forever, slowing down.	Infinite	Expands forever. No recollapse.	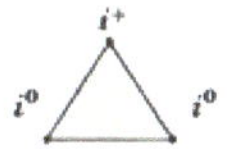
−1 (Open)	Matter	$\Omega_0 < 1$	Expands forever, slowing down.	Infinite	Expands forever, curvature dominates.	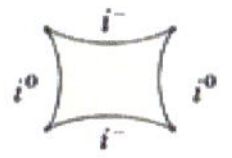

With $\Lambda > 0$, even closed ($k = +1$) universes can expand forever if Λ is large enough.

Redshift and Distance

Redshift of light emitted at t_e and observed at t_0:

$$1 + z = \frac{a(t_0)}{a(t_e)}$$

Comoving distance:

$$\chi = \int_{t_e}^{t_0} \frac{c\,dt}{a(t)} = \int_{a(t_e)}^{a(t_0)} \frac{c\,da}{a^2 H(a)}$$

Proper distance today:

$$D(t_0) = a(t_0) S_k(\chi)$$

where

$$S_k(\chi) = \begin{cases} \sin\chi & (k = +1) \\ \chi & (k = 0) \\ \sinh\chi & (k = -1) \end{cases}$$

Scale Factor Examples ($\Lambda = 0$)

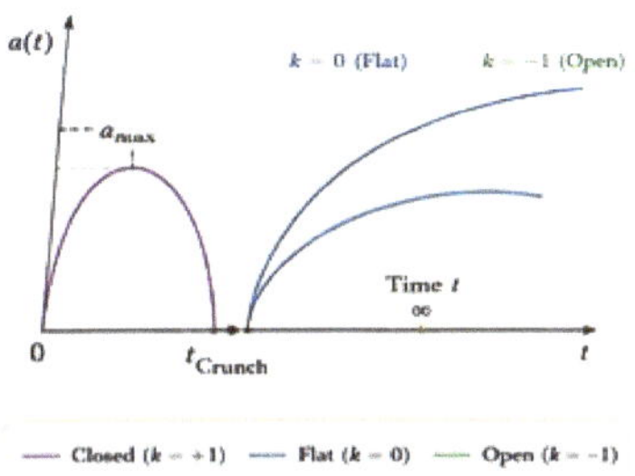

Energy Content Today (Planck 2018)

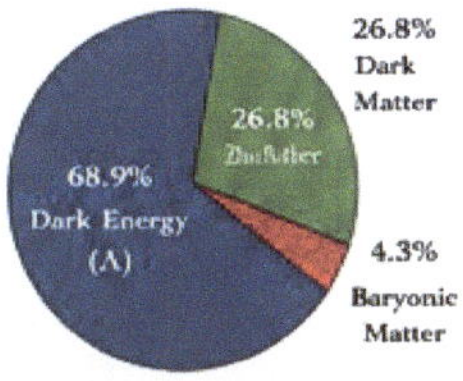

$\Omega_m \approx 0.315$, $\Omega_\Lambda \approx 0.685$, $\Omega_k \approx 0.001$

$(\Omega_m + \Omega_\Lambda + \Omega_k = 1)$

Cosmic History (Qualitative)

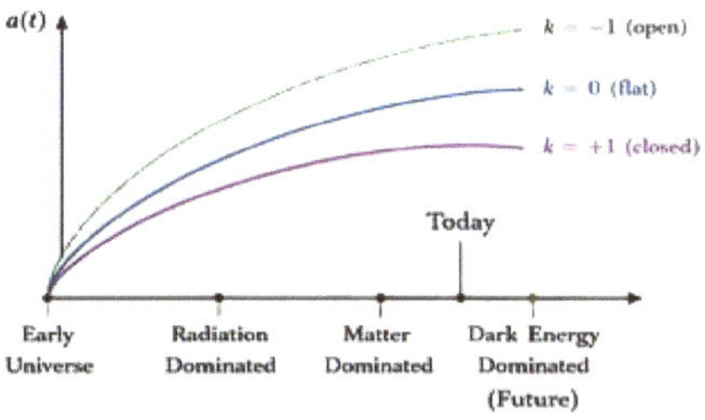

Special Notes

- Einstein–de Sitter universe on the largest $\Omega_m = 1$, $\Omega_\Lambda = 0$, k few parameters.
- Milne geometry, matter, and expansion in a simple framework.
- Explains cosmic microwave background isotropy and large-scale structure.
- Foundation for studying structure formation, dark energy, and inflation.
- Helps us understand the origin, evolution, and ultimate fate of our universe.

Key Takeaway

FLRW solutions show how curvature (k), matter (ρ), pressure (p), and cosmological constant (Λ) shape the expansion history and destiny of the universe—from the Big Bang to infinite expansion, recollapse, or accelerated growth.

17 PENROSE DIAGRAM OF SCHWARZSCHILD–DE SITTER SPACETIME

The Schwarzschild–de Sitter (SdS) spacetime describes a static, spherically symmetric black hole in a universe with a positive cosmological constant ($\Lambda > 0$).

KEY FEATURES

- Static, spherically symmetric solution with $\Lambda > 0$.
- Two positive real horizons exist when $0 < 9M^2\Lambda < 1$.
- Contains a black hole and a cosmological horizon.
- Globally maximally extended spacetime has 9 distinct regions.
- Singularities at $r = 0$ (timelike).

THE METRIC

The SdS metric in static coordinates is

$$ds^2 = -f(r)dt^2 + \frac{dr^2}{f(r)} + r^2 d\Omega^2$$

where

$$f(r) = 1 - \frac{2M}{r} - \frac{\Lambda r^2}{3}$$

The positive roots of $f(r) = 0$ give the horizon radii:

$$r_n < r_c$$

(black hole < cosmological).

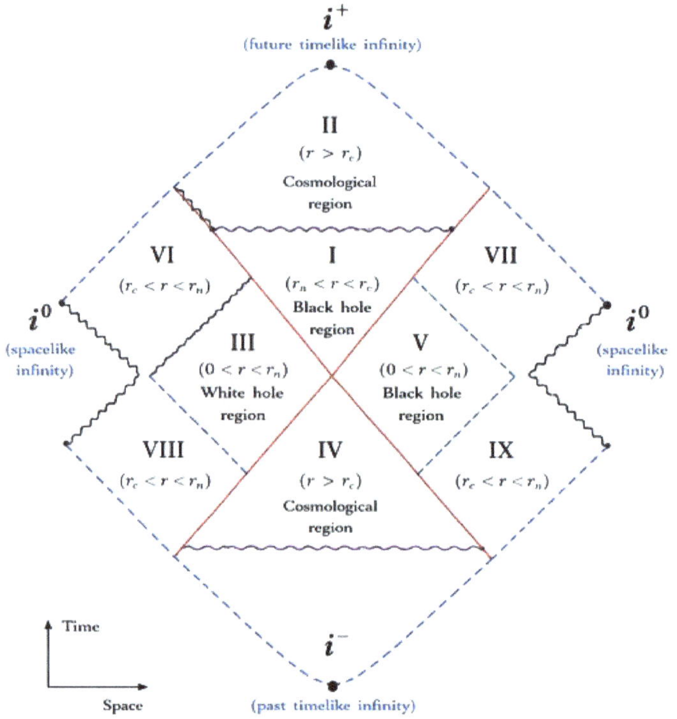

HORIZONS

- Black hole horizon ($r = r_n$)
- Cosmological horizon ($r = r_c$)
- Cauchy horizon for inner regions.
- Timelike singularity ($r = 0$).

Ordering:

$$0 < r_n < r_c < \infty$$

REGIONS

I : External black hole region (between r_n and r_c).

II : Cosmological regions ($r > r_c$).

III : White hole interior ($r < r_n$, past origin).

V : Black hole interior ($r < r_n$, future singularity).

VI–IX: Other asymptotic regions related by analytic extension.

- Black hole horizon ($r = r_n$)
- Cosmological horizon ($r = r_c$)
- Timelike singularity ($r = 0$)
- Spacelike singularity ($r = 0$ at i^0)

- Cosmic observers in region I see a black hole bounded by r_n and a cosmological horizon at r_c.
- Regions II and IV represent expanding (or contracting) de Sitter universes beyond the cosmological horizon.
- The Penrose diagram is periodic in the vertical (time) direction.

EXISTENCE OF HORIZONS

Horizons exist when $0 < 9M^2\Lambda < 1$. The critical (Nariai) limit is $9M^2\Lambda = 1$, where $r_n = r_c = \sqrt{\frac{1}{\Lambda}}\cdot\sqrt{3}$.

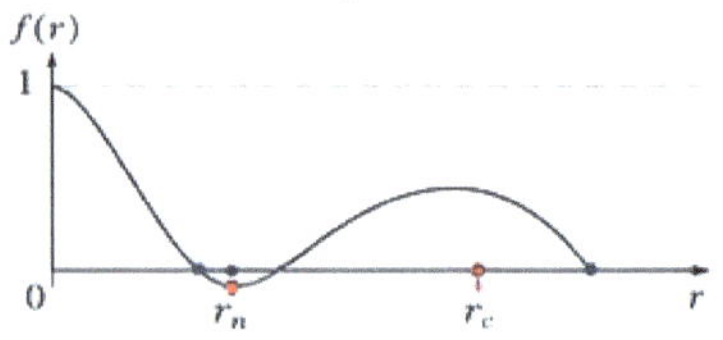

CAUSAL STRUCTURE

No signal from beyond r_c can reach the black hole or the central singularity.

$R = 2M$ is not a horizon here ($f(2M) > 0$).

All timelike curves in regions II and IV approach i^+ or i^- as proper time goes to $\pm\infty$.

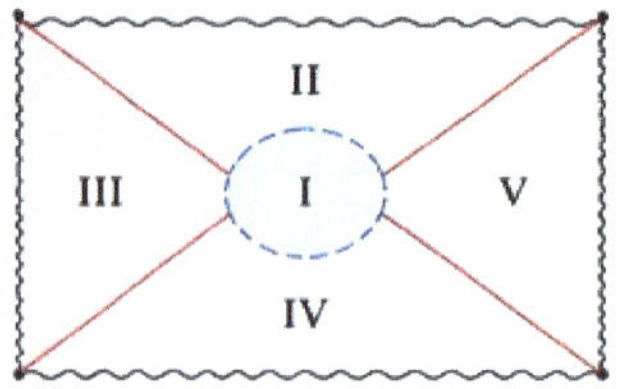

PHYSICS AND IMPLICATIONS

- Describes a black hole embedded in a universe with positive Λ (e.g., our accelerating universe).
- Cosmological horizon introduces an ultimate causal boundary.
- Thermal properties:
  - Black hole temperature $T_b = \frac{\kappa_n}{2\pi}$
  - Cosmological temperature $T_c = \frac{\kappa_c}{2\pi}$
- May lead to rich thermodynamics and possible interactions between the two horizons.
- Important in studies of black hole thermodynamics, quantum fields in curved spacetime, and holography.

SCALE FACTOR $a(t)$ ($\Lambda > 0$)

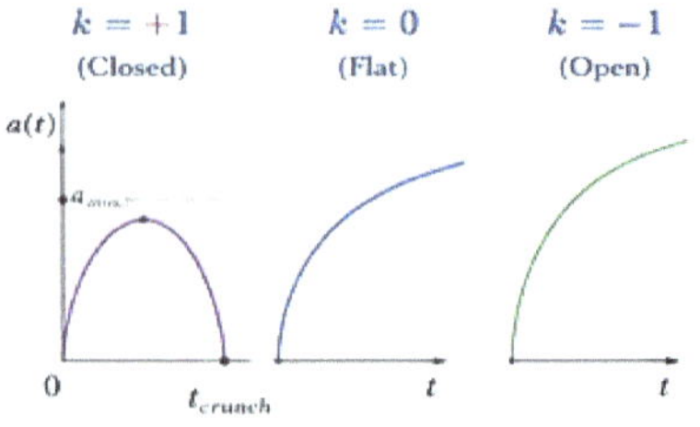

— Closed ($k = +1$) — Flat ($k = 0$) — Open ($k = -1$)

ENERGY CONTENT TODAY (PLANCK 2018)

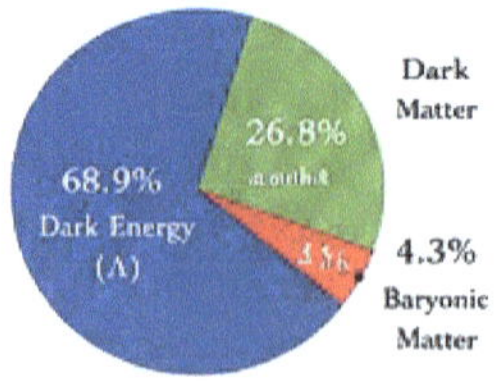

$\Omega_\Lambda \approx 0.69$, $\Omega_{DM} \approx 0.27$, $\Omega_b \approx 0.04$

$\Omega_{total} \approx 1.00$ (flat universe)

COSMIC HISTORY (QUALITATIVE)

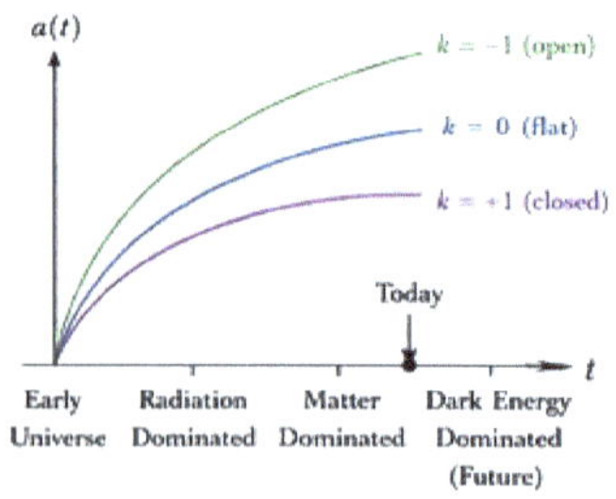

WHY FLRW MATTERS

- Describes the universe on the largest scales with only a few parameters.
- Links geometry, matter, and expansion in a simple framework.
- Explains cosmic microwave background isotropy and large-scale structure.
- Foundation for studying structure formation, dark energy, and inflation.
- Helps us understand the origin, evolution, and ultimate fate of our universe.

KEY TAKEAWAY

FLRW solutions show how curvature (k), matter (ρ), pressure (p), and the cosmological constant (Λ) shape the expansion history and destiny of the universe—from the Big Bang to infinite expansion, recollapse, or accelerated growth.

18

PENROSE DIAGRAM OF KERR–de SITTER SPACETIME

Rotating Black Hole in a Universe with Positive Cosmological Constant ($\Lambda > 0$)

Kerr–de Sitter (KdS) spacetime describes a rotating black hole embedded in a de Sitter universe.
It has two horizons: the black hole event horizon (r_+) and the cosmological horizon (r_c).

KEY FEATURES

- Solution of Einstein's equations with mass M, angular momentum a, and $\Lambda > 0$.
- Two real positive horizons when $0 < a < \Lambda_{max}(a)$: $0 < r_- < r_+ < r_c$.
- Contains an ergosphere between r_- and r_+.
- Globally regular (no naked singularity): ring singularity at $r = 0$.
- Asymptotically approaches de Sitter spacetime.

HORIZON RADII

The horizons are roots of $\Delta_r = 0$, where

$$\Delta_r = (r^2 + a^2)\left(1 - \frac{\Lambda r^2}{3}\right) - 2Mr$$

For $0 < a < \Lambda_{max}(a)$, there are three real roots:

$$0 < r_- < r_+ < r_c$$

r_- : Cauchy horizon (inner)
r_+ : Event horizon (outer)
r_c : Cosmological horizon

ERGOSPHERE

The ergosurface is given by

$$g_{tt} = 0 \Rightarrow \Delta_r - a^2 \sin^2\theta = 0$$

It lies outside the event horizon r_+ and touches it at the poles ($\theta = 0, \pi$).

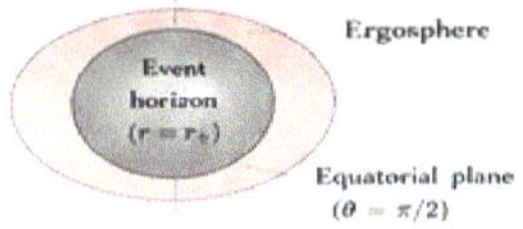

PENROSE DIAGRAM (MAXIMAL EXTENSION)

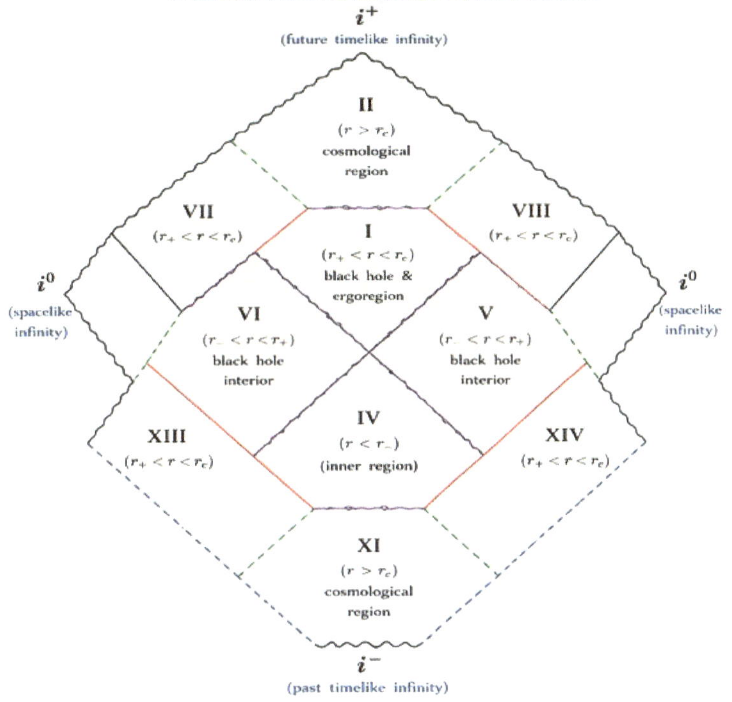

Red solid : Event horizon ($r = r_+$)
Green dashed : Cosmological horizon ($r = r_c$)
Purple wavy : Ring (curvature) singularity ($r = 0$, $\theta = \pi/2$)
Black wavy : Timelike infinity ($i^{\pm}$)

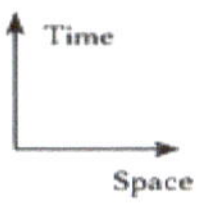

BOUNDARIES & INFINITIES

- i^+ : future timelike infinity (future/past).
- i^0 : spacelike infinity.
- $\mathcal{J}^+$: future cosmological infinity (spacelike).
- $\mathcal{J}^-$: past cosmological infinity (spacelike).
- $\mathcal{J}^0$: spacelike infinity.

No null infinity ($\mathcal{J}^{\pm}$) exists because $\Lambda > 0$.

REGIONS

I : Black hole exterior between r_+ and r_c. Contains ergoregion.
II, XI : Cosmological regions beyond r_c.
III : White hole interior ($r_- < r_+$, past origin).
V, VI : Black hole interiors between r_- and r_+.
IV : Deep interior ($r < r_-$), reaches the ring singularity.
VII, VIII, XIII, XIV : Other asymptotic regions related by analytic extension.

IMPORTANT FACTS

- Bounded by two horizons: black hole (r_+) and cosmological (r_c).
- Causal curves from I cannot reach $r > r_c$.
- Penrose diagram is periodic in the vertical (time) direction.
- Reduces to Kerr for $\Lambda \to 0$ ($r_c \to \infty$).
- Reduces to Schwarzschild–de Sitter for $a \to 0$.

HORIZON STRUCTURE

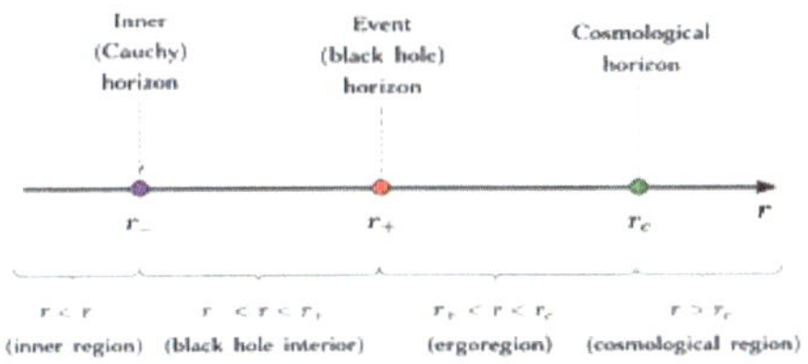

EXTREME & NARIAI LIMITS

There exists a maximum $\Lambda_{max}(a)$ such that $0 < a < \Lambda_{max}(a)$ for three real horizons.

- $\Lambda \to \Lambda_{max}(a)$: Nariai limit ($r_+ \to r_c$) $\to$ black hole and cosmological horizons coincide.
- $a \to 0$: Schwarzschild–de Sitter.
- $a \to \Lambda_{max}(a)$: extreme ($r_- \to r_+$).
- $a \to 0$ and $\Lambda \to 0$: Minkowski.

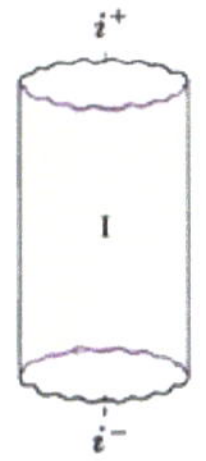

ANGULAR MOMENTUM EFFECTS

- Rotation ($a \neq 0$) creates the ergosphere, allowing energy extraction (Penrose process).
- Frame dragging occurs in regions $r_- < r < r_+$ and outside.

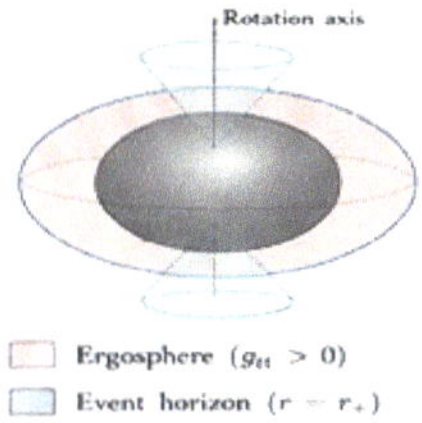

METRIC (BOYER–LINDBLAD COORDINATES)

$$ds^2 = -\left(1 - \frac{2Mr}{\Sigma} - \frac{\Lambda r^2}{3}\right) dt^2 - \frac{4Mar\sin^2\theta}{\Sigma}\, dt\, d\phi + \frac{\Sigma}{\Delta_r} dr^2 + \frac{\Sigma}{\Delta_\theta} d\theta^2 + \frac{A \sin^2\theta}{\Sigma} d\phi^2$$

where

$$\Sigma = r^2 + a^2\cos^2\theta, \quad \Delta_r = (r^2 + a^2)\left(1 - \frac{\Lambda r^2}{3}\right) - 2Mr$$

$$\Delta_\theta = 1 + \frac{\Lambda a^2}{3}\cos^2\theta, \quad \Sigma = (r^2 + a^2)^2 - a^2\Delta_r \sin^2\theta$$

PHYSICAL INTERPRETATION

- Region I describes our universe between the black hole horizon and the cosmological horizon.
- The universe is causally isolated by the cosmological horizon.
- Thermal radiation arises from both horizons:
  - Black hole temperature $T_b = \frac{\kappa_+}{2\pi}$
  - Cosmological temperature $T_c = \frac{\kappa_c}{2\pi}$
- Rich structure: horizons, ergospheres, frame dragging, and global periodicity.

SURFACE GRAVITY & TEMPERATURES

Surface gravities at the horizons:

$$\kappa_+ = \frac{\Delta_r'(r_+)}{2(r_+^2 + a^2)}, \quad \kappa_c = \kappa(r_c)$$

$\kappa_- = \; < 0$ (inner horizon)

Hawking–Gibbons temperatures:

$$T_h = \frac{\kappa_+}{2\pi} \quad \text{(black hole horizon)}$$

$$T_c = \frac{\kappa_c}{2\pi} \quad \text{(cosmological horizon)}$$

Generally, $T_h \neq T_c$.

SCALE FACTOR EXAMPLES ($\Lambda > 0$)

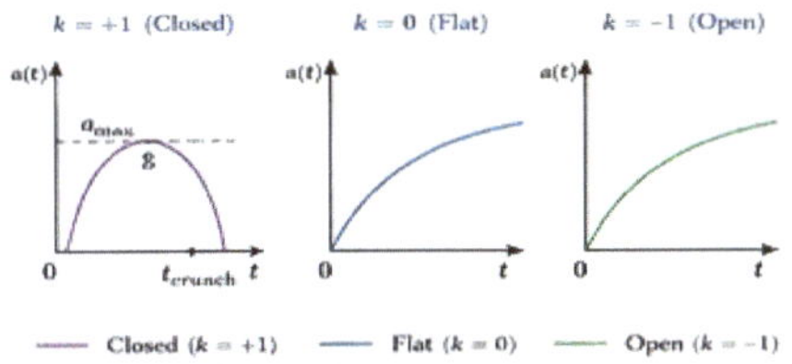

ENERGY CONTENT TODAY

(PLANCK 2018)

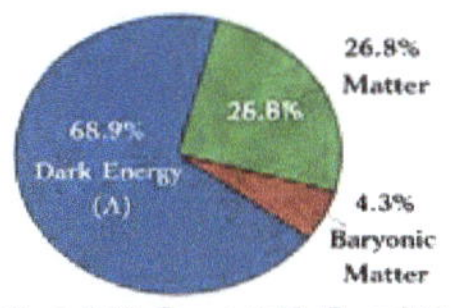

$\Omega_\Lambda \approx 0.69$, $\Omega_{DM} \approx 0.27$, $\Omega_b \approx 0.04$
$\Omega_{total} \approx 1.00$ (flat universe)

COSMIC HISTORY (QUALITATIVE)

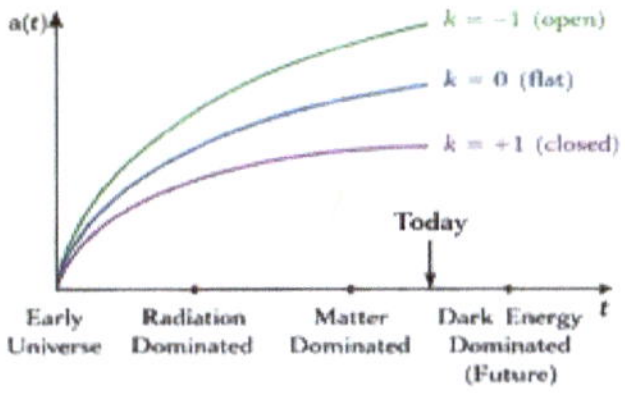

WHY FLRW MATTERS

- Describes the universe on the largest scales with only a few parameters.
- Links geometry, matter, and expansion in a simple framework.
- Explains cosmic microwave background isotropy and large-scale structure.
- Foundation for studying structure formation, dark energy, and inflation.
- Helps us understand the origin, evolution, and ultimate fate of our universe.

KEY TAKEAWAY

The Kerr–de Sitter Penrose diagram shows a rotating black hole in a de Sitter universe with ts and a reich causal structure. It captures the interplay between, rotation, and cosmic, causally tmmlly bounded universe.

19 PENROSE DIAGRAM OF THE BTZ BLACK HOLE

(2+1 DIMENSIONS, $\Lambda < 0$)

The BTZ black hole is the (2+1)-dimensional solution of Einstein's equations with negative cosmological constant. It is locally AdS_3 with a black hole horizon and a timelike singularity.

BTZ BLACK HOLE

- Solution to Einstein's equations with $\Lambda = -\frac{1}{\ell^2} < 0$ in (2+1) D.
- Characterized by mass M and angular momentum J ($|J| \leq M\ell$).
- Event horizons at $r = r_\pm$.

$$r_\pm = \ell\sqrt{\frac{M}{2}}\left(1 \pm \sqrt{1 - \frac{J^2}{M^2\ell^2}}\right)$$

- $r_+ = 0$ is a timelike curvature singularity.
- Asymptotically AdS_3.

THE METRIC

In Schwarzschild-like coordinates (t, r, ϕ):

$$ds^2 = -N^2(r)\,dt^2 + N^{-2}(r)\,dr^2 + r^2\,(d\phi + N^\phi(r)\,dt)^2$$

where

$$N^2(r) = -M + \frac{r^2}{\ell^2} + \frac{J^2}{4r^2}$$

$$N^\phi(r) = -\frac{J}{2r^2}$$

Horizon(s): $N^2(r_+) = 0$ (outer)
$N^2(r_-) = 0$ (inner)

PENROSE DIAGRAM (MAXIMAL EXTENSION)

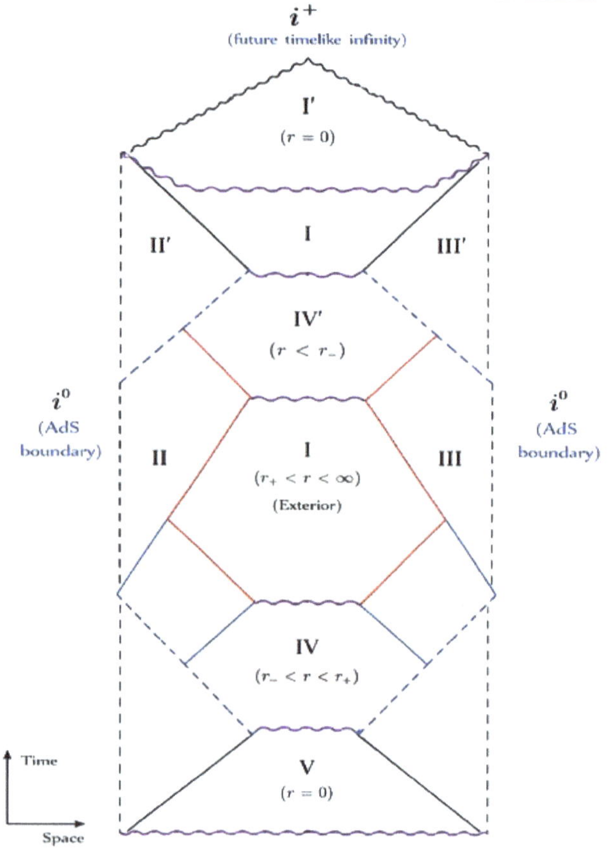

Timelike Singularity ($r = 0$)

Red solid	:	Event horizon ($r = r_+$)
Green dashed	:	Cosmological horizon ($r = r_c$)
Purple wavy	:	Ring (curvature) singularity ($r = 0, \theta = \frac{\pi}{2}$)
Black wavy	:	Timelike infinity ($i^\pm$)
Blue dashed	:	Cauchy horizon ($r = r_-$)
Black solid	:	Conformal boundary ($r \to \infty$)
Gray dashed	:	Null surface

LEGEND

—	Red solid	Event horizon ($r = r_+$)
- - -	Blue dashed	Cauchy horizon ($r = r_-$)
~~	Purple wavy	Timelike singularity ($r = 0$)
- - -	Black dashed	Conformal boundary ($r \to \infty$) = AdS_3
—	Gray dashed	Null surface

REGIONS

- I : Exterior black hole region between r_+ and r_c. Contains ergoregion.
- II, XI : Cosmological regions beyond r_c.
- III : White hole interior ($r < r_+$, past origin).
- V, VI : Black hole interiors between r_- and r_+.
- IV : Region between inner and outer horizons ($r_- < r < r_+$).
- IV′ : Inside Cauchy horizon ($r < r_-$).
- I′, II′, III′ : Future exterior regions ($r > r_c$).

BOUNDARIES & INFINITIES

- i^+ : future timelike infinity (future/past).
- i^0 : spacelike infinity.
- $\mathcal{J}^+$: future conformal infinity (spacelike).
- $\mathcal{J}^-$: past conformal infinity (spacelike).

No null infinity ($\mathcal{J}^\pm$) exists because $\Lambda < 0$.

GLOBAL STRUCTURE

- The singularity at $r = 0$ is timelike.
- There is an event horizon at $r = r_+$.
- There is a Cauchy horizon at $r = r_-$ (if $J \neq 0$).
- The spacetime is periodic in ϕ with $0 \leq \phi < 2\pi$.
- The diagram is the same for all $M > 0$ and $|J| \leq M\ell$.

RADIAL STRUCTURE

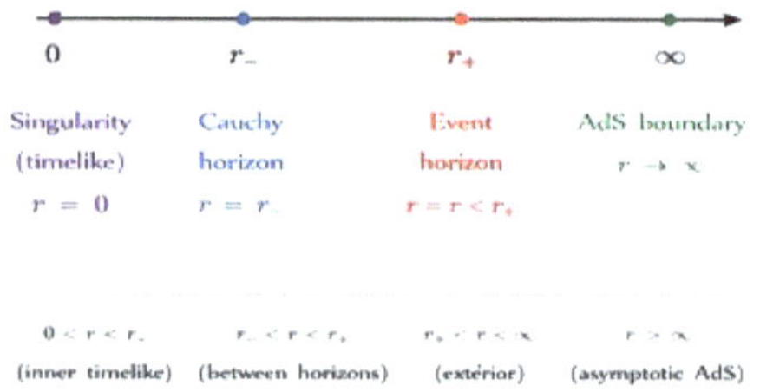

CIRCULAR ORBITS (QUALITATIVE)

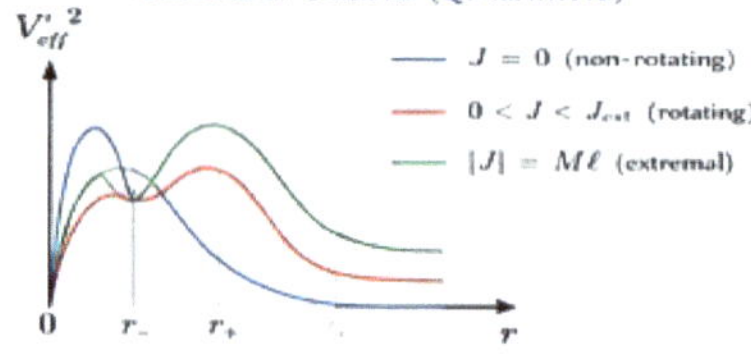

Stable circular orbits exist only for $r > r_+$.

CIMPORMAL BOUNDARY (AdS_3)

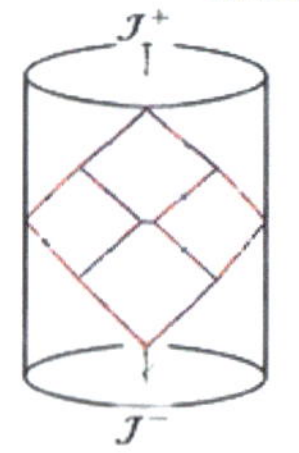

- The Penrose diagram can be obtained by unwrapping the cylinder along ϕ.
- Points on the vertical edges are identified ($\phi \sim \phi + 2\pi$).

EXTREME CASE ($J = \pm M\ell$)

Inner and outer horizons coincide:

$$r_+ = r_- = r_e = \ell\sqrt{\frac{M}{2}}$$

- Cauchy horizon disappears.
- Penrose diagram reduces to a simpler diamond with a single horizon.

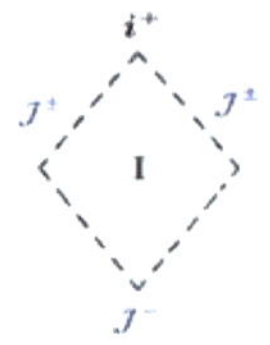

PHYSICAL PROPERTIES

- Mass: $M \geq 0$
- Angular momentum: $|J| \leq M\ell$
- Horizon lengths: $2\pi r_\pm$
- Bekenstein–Hawking entropy: $S = \frac{2\pi r_+}{4G_3} = \frac{\pi r_+}{2G_3}$.
- Temperature: $T_H = \frac{r_+^2 - r_-^2}{2\pi\ell^2 r_+}$.
- Angular velocity at horizon: $\Omega_H = \frac{r_-}{\ell r_+}$.

NOTES

- No gravitational waves in (2+1) D.
- BTZ black hole is a laboratory for quantum gravity, AdS/CFT, and near-horizon physics.
- The diagram is periodic in the vertical direction ($t \sim t + \beta$ for Euclidean time).
- The BTZ spacetime is globally AdS_3 but with a black hole carved out by identifications.

KEY TAKEAWAY

The BTZ Penrose diagram reveals a rich causal structure with an outer event horizon, an inner Cauchy horizon (for $J \neq 0$), and a timelike singularity at $r = 0$, all within asymptotically AdS_3 spacetime.

20 PENROSE DIAGRAM OF VAIDYA SPACETIME

Spherically Symmetric Radiating Black Hole

The Vaidya spacetime describes the gravitational collapse (or evaporation) of null dust.
The mass $M(v)$ changes with the advanced time v, representing ingoing (or outgoing) radiation.

THE VAIDYA SPACETIME

Spherically symmetric solution of Einstein's equations for null dust. In advanced Eddington–Finkelstein coordinates (v, r, θ, ϕ):

$$ds^2 = -\left(1 - \frac{2M(v)}{r}\right) dv^2 + 2\, dv\, dr + r^2\, d\Omega^2$$

where

- $v = t + r^*$: advanced null time
- r : areal radius
- $M(v)$: mass function
- $d\Omega^2 = d\theta^2 + \sin^2\theta\, d\phi^2$
- $T_{ab} = \frac{1}{4\pi r^2} \frac{dM}{dv} l_a l_b$

$l^a = \partial_r$: ingoing radial null vector

PHYSICAL MEANING

$$M'(v) = \frac{dM}{dv}$$

- $M'(v) > 0$: ingoing radiation (collapse)
- $M'(v) < 0$: outgoing radiation (evaporation)
- $M'(v) = 0$: reduces to Schwarzschild spacetime

APPARENT HORIZON

The apparent horizon is located at $r = 2M(v)$.

It is marginally trapped:

$$\theta_{(\ell)} = 0, \quad \theta_{(n)} < 0$$

It is spacelike if $M' > 0$ (collapse) and timelike if $M' < 0$ (evaporation).

PENROSE DIAGRAM (DYNAMIC CASE: $M'(v) \neq 0$)

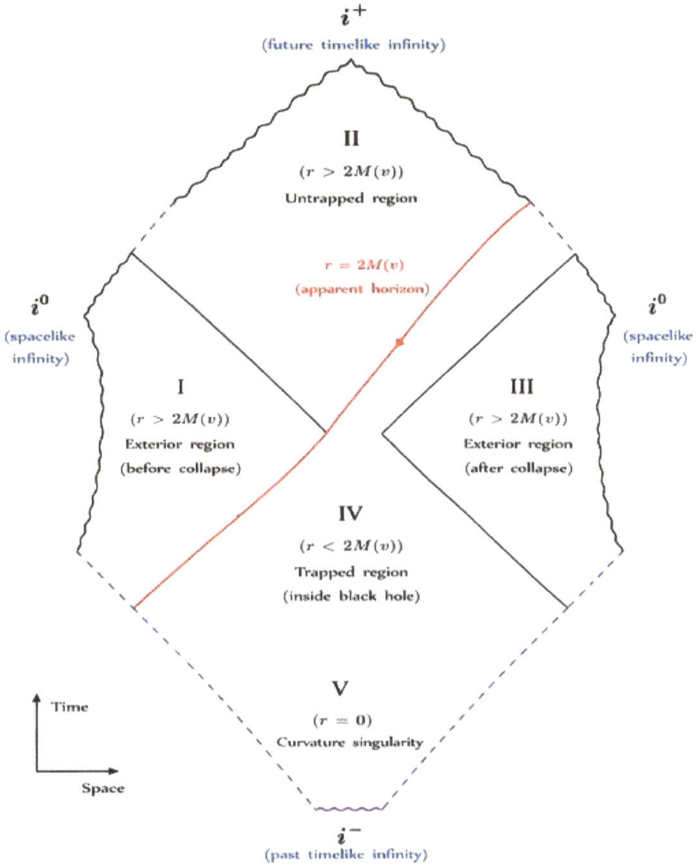

Red solid	:	Apparent horizon $r = 2M(v)$
Purple wavy	:	Curvature singularity $r = 0$ (timelike)
Blue dashed	:	Null infinity (boundary)
Black wavy	:	Timelike infinity
Black solid	:	Null surface (constant v)

LEGEND

- Red solid line : Apparent horizon $r = 2M(v)$
- Purple wavy line : Curvature singularity $r = 0$
- Blue dashed line : Null infinity (boundary)
- Black wavy line : Timelike infinity
- Black solid line : Null surface (constant v)

REGIONS

I	:	Exterior region before collapse ($r > 2M(v)$)
II	:	Untrapped region outside singularity and horizon ($r > 2M(v)$)
III	:	Exterior region after collapse ($r > 2M(v)$)
IV	:	Trapped region inside black hole ($r < 2M(v)$)
V	:	Singularity region ($r = 0$)

BOUNDARIES & INFINITIES

i^+	:	future timelike infinity
i^-	:	past timelike infinity
i^0	:	spacelike infinity
$r = 0$	:	curvature singularity (timelike)
Dashed lines	:	null infinity
Solid null lines	:	surfaces of constant v

KEY PROPERTIES

- ★ Spherical symmetry with null dust.
- ★ Generalizes Schwarzschild by allowing time-dependent mass $M(v)$.
- ★ Describes gravitational collapse or evaporation.
- ★ Apparent horizon $r = 2M(v)$ is marginally trapped.
- ★ Future development depends on the behavior of $M(v)$.

RADIAL NULL GEODESICS

$$0 = -\left(1 - \frac{2M(v)}{r}\right) dv^2 + 2\, dv\, dr$$

$$\Rightarrow \frac{dr}{dv} = \frac{1}{2}\left(1 - \frac{2M(v)}{r}\right)$$

- Outgoing null rays: $\frac{dr}{dv} > 0$
- Ingoing null rays: $\frac{dr}{dv} < 0$
- At $r = 2M(v)$, outgoing rays momentarily hover (apparent horizon).

COLLAPSING VS EVAPORATING CASES

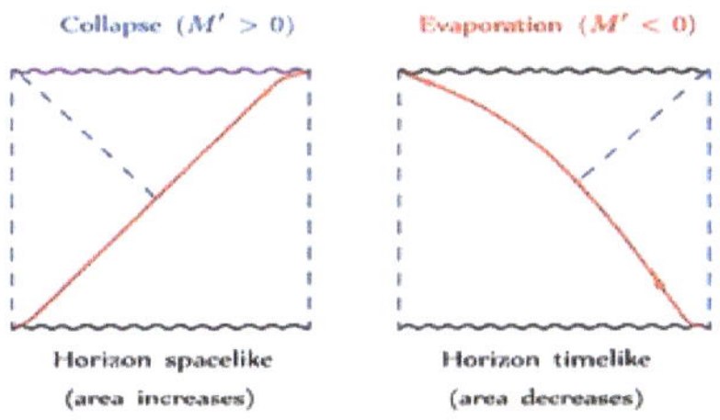

PHYSICAL PICTURE

- Before collapse (Region I): spacetime nearly flat.
- During collapse: mass increases, horizons form.
- After collapse (Region III): approaches Schwarzschild.
- Inside (Region IV): all matter and light fall into the singularity at $r = 0$.

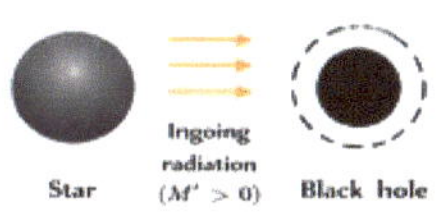

CAUSAL STRUCTURE SUMMARY

Quantity	Expression / Condition	Meaning
Mass function	$M(v)$	Time-dependent mass
Energy flux	$T_{vv} = \frac{1}{4\pi r^2} \frac{dM}{dv}$	Ingoing (>0) or outgoing (<0) radiation
Apparent horizon	$r = 2M(v)$	Boundary of trapped region
Singularity	$r = 0$	Timelike curvature singularity

IMPORTANT EQUATIONS

- Metric:
$$ds^2 = -\left(1 - \frac{2M(v)}{r}\right) dv^2 + 2\, dv\, dr + r^2\, d\Omega^2$$
- Energy–momentum tensor:
$$T_{ab} = \frac{1}{4\pi r^2} \frac{dM}{dv} l_a l_b$$
- Apparent horizon: $r = 2M(v)$
- Radial null geodesics:
$$\frac{dr}{dv} = \frac{1}{2}\left(1 - \frac{2M(v)}{r}\right)$$

NOTES

- The Penrose diagram is conformal: it shows causal structure, not distances.
- It is valid for any function $M(v)$.
- For evaporation ($M' < 0$), the apparent horizon becomes timelike.
- For collapse ($M' > 0$), the apparent horizon becomes spacelike.

KEY TAKEAWAY

The Vaidya Penrose diagram captures the dynamical formation (or evaporation) of a black hole through null radiation, with a time-dependent horizon separating trapped and untrapped regions.

SUMMARY OF PENROSE DIAGRAMS

A COMPARATIVE OVERVIEW OF KEY SPACETIMES

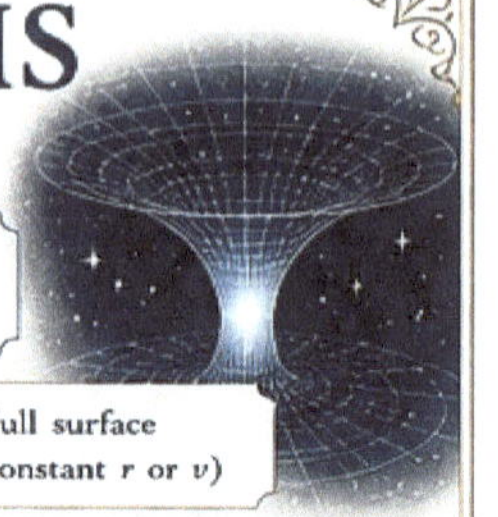

Penrose diagrams provide a powerful way to represent the causal structure of spacetimes in a compact 2D plot. They preserve the light-cone structure, showing how events are connected by timelike, null, or spacelike paths.

Null infinity (boundary) | Curvature singularity ($r = 0$) | Event horizon (or apparent horizon) | Timelike infinity (i^+, i^-) | Null surface (constant r or v)

1. SCHWARZSCHILD BLACK HOLE

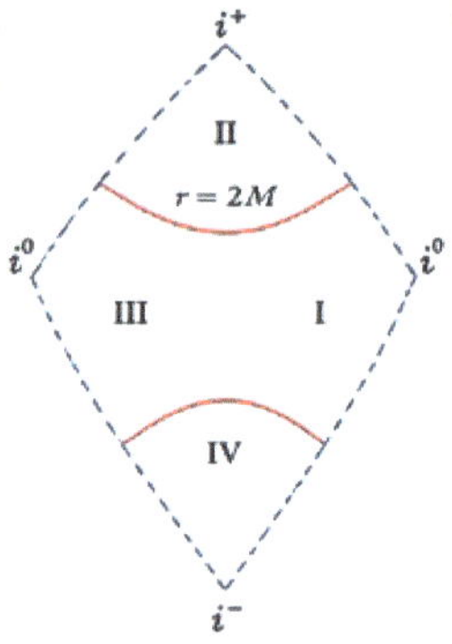

- Static, spherically symmetric black hole ($\Lambda = 0$).
- Event horizon at $r = 2M$.
- Singularity at $r = 0$.
- Region I: exterior; Region IV: interior (trapped).
- i^+ and i^- are timelike infinities.

2. SCHWARZSCHILD–DE SITTER (SdS)

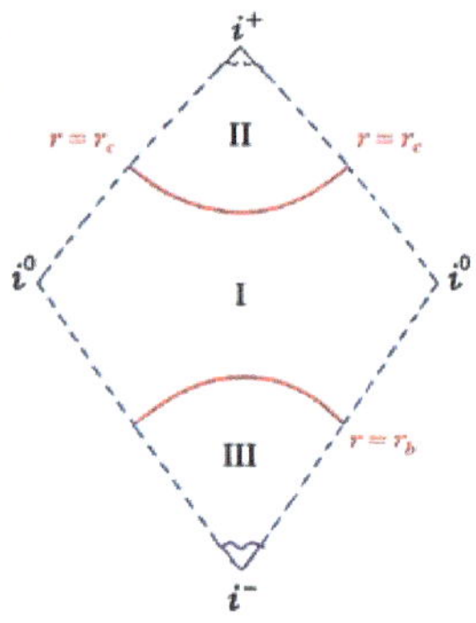

- $\Lambda > 0$ introduces a cosmological horizon r_c.
- Two horizons: black hole (r_b) and cosmological (r_c).
- Static region (I) between the two horizons.
- Singularity at $r = 0$.
- i^+ is spacelike.

3. KERR BLACK HOLE

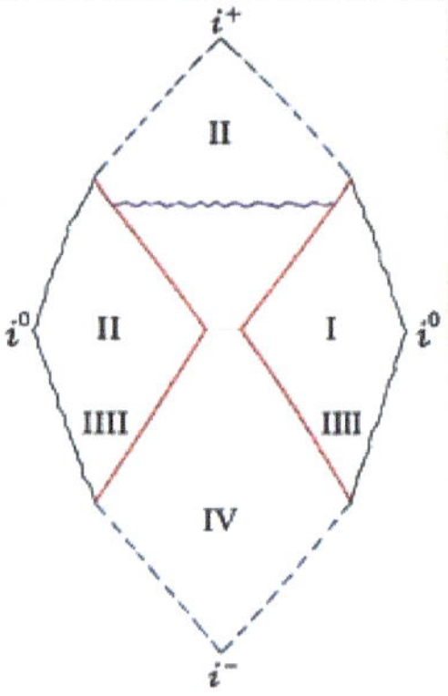

- Rotating black hole ($\Lambda = 0$).
- Two horizons: outer (r_+) and inner (r_-).
- Ergosphere lies outside r_+.
- Singularity is a ring ($r = 0$, $\theta = \pi/2$).
- Causal structure richer due to frame dragging.

4. KERR–DE SITTER (KdS)

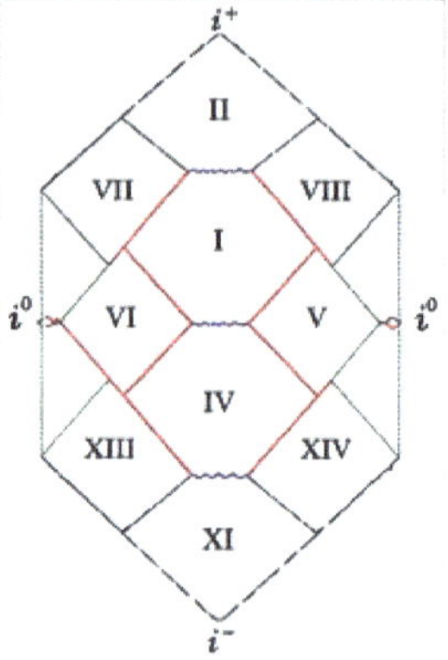

- Rotating black hole in de Sitter universe ($\Lambda > 0$).
- Four horizons: Cauchy (r_-), event (r_+), cosmological (r_c), and inner cosmological.
- Static region (I) between r_+ and r_c.
- More regions due to both rotation and $\Lambda > 0$.

5. BTZ BLACK HOLE (2+1 D)

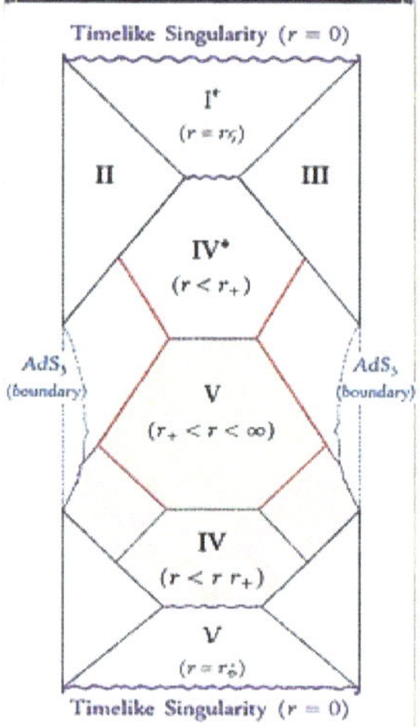

- 2+1 dimensional black hole with $\Lambda < 0$ (AdS_3).
- One event horizon at r_+.
- No local gravitational degrees of freedom.
- Timelike boundary represents AdS_3 infinity.
- Singularity at $r = 0$.

6. VAIDYA SPACETIME (RADIATING / COLLAPSING)

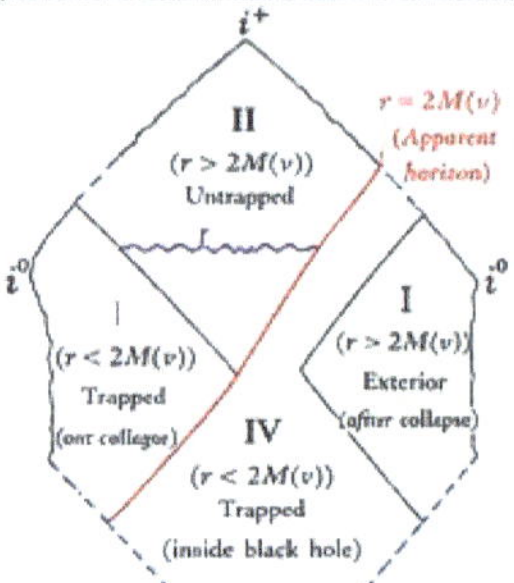

- Describes gravitational collapse $M(v) \geq 0$ or evaporation $M(v) < 0$.
- Apparent horizon: $r = 2M(v)$ is time-dependent.
- No global event horizon in general.
- Reduces to Schwarzschild for constant M.

7. MINKOWSKI (FLAT SPACETIME)

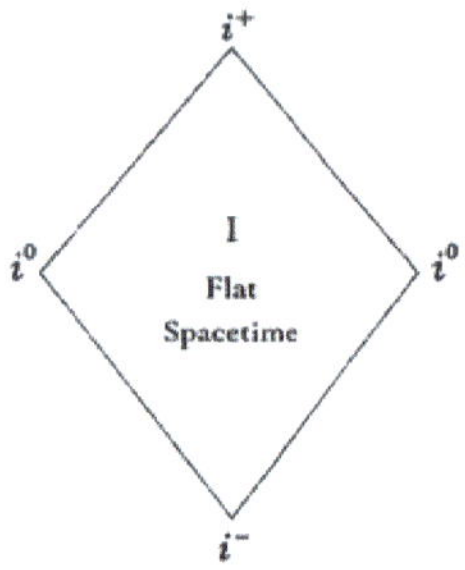

- No gravity (η_{ab}).
- Light rays at 45°.
- Entire spacetime is regular.
- Baseline for comparison.

8. RINDLER (UNIFORMLY ACCELERATED)

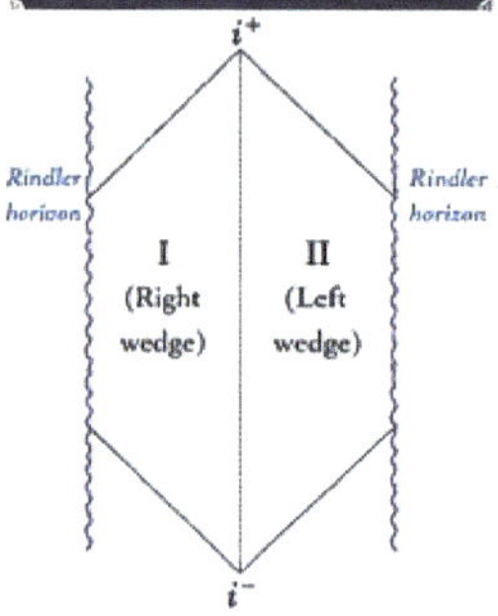

- Coordinates for uniformly accelerated observers.
- Rindler horizons are coordinate (not curvature) singularities.
- Minkowski spacetime viewed from accelerated frame.

9. EXTREMAL REISSNER–NORDSTRÖM

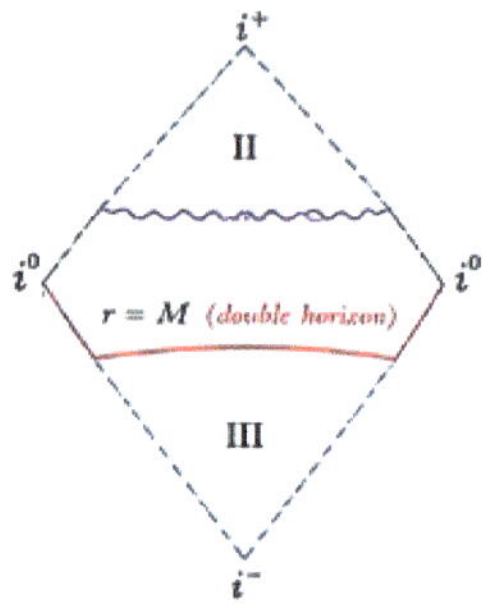

- Extremal case: $Q = M$.
- Degenerate horizon at $r = M$.
- Infinite throat near horizon.
- Singularity at $r = 0$.

10. ANTI–DE SITTER (GLOBAL AdS)

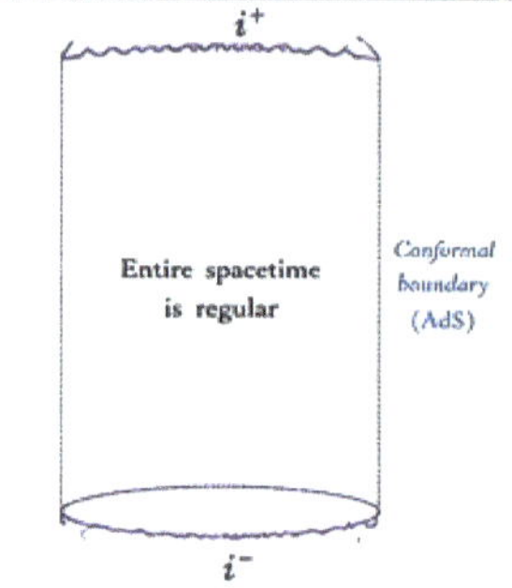

- $\Lambda < 0$, no black hole.
- Timelike infinity; signals reflect off boundary and return.
- Important in AdS/CFT correspondence.

KEY TAKEAWAYS

- ★ Penrose diagrams capture causal structure, not distances.
- ★ Horizons (event or apparent) separate regions of causal (in)accessibility.
- ★ Singularity lines represent physical or coordinate pathologies.
- ★ $\Lambda > 0$ adds cosmological horizons; $\Lambda < 0$ leads to conformal boundaries.
- ★ Rotation (Kerr) and radiation (Vaidya) enrich causal structure significantly.

COMMON FEATURES

Feature	Schwarzschild	SdS	Kerr	KdS	BTZ	Vaidya
Λ	0	> 0	0	> 0	< 0	0
Rotation	No	No	Yes	Yes	No	No
Horizons	1	2	2	4	1	1 (time-dependent)
Singularity	$r = 0$	$r = 0$	Ring	Ring	$r = 0$	$r = 0$ (center)
Infinity	Timelike	Spacelike	Timelike	Spacelike	Timelike	Mixed
Asymptotics	Flat	dS	Flat	dS/AdS	AdS_3	Dynamical

READING A PENROSE DIAGRAM

1. Identify infinities (i^+, i^-, i^0, $\mathscr{I}^+$).
2. Locate horizons (event / apparent).
3. Note singularities (curvature or coordinate).
4. Determine regions (exterior, trapped, etc.).
5. Follow null lines at 45° to understand causal connections.

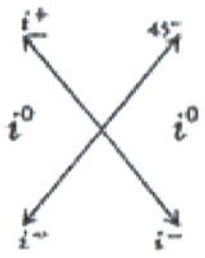

IN ESSENCE

Penrose diagrams are maps of possibility and limitation—showing what can influence what, and how the universe is woven by light, gravity, and fate.

www.ingramcontent.com/pod-product-compliance
Lightning Source LLC
LaVergne TN
LVHW070625061126
851857LV00002B/15